Modern Policing Using ArcGIS® Pro

MODERN POLICING

Using ArcGIS® Pro

ERIC L. PIZA &
JONAS H. BAUGHMAN

Esri Press
REDLANDS | CALIFORNIA

Esri Press, 380 New York Street, Redlands, California 92373-8100
Copyright © 2021 Esri
All rights reserved.
25 24 23 22   4 5 6 7 8 9 10

ISBN: 9781589485976
Library of Congress Control Number: 2020952370

Contents

Preface

The authors base this book around two core principles: (1) crime analysis has provided the engine for the evolution of modern policing, and (2) the work of crime analysts is critical in maximizing the effect of contemporary crime prevention strategies. This book further recognizes the key role that crime mapping and spatial analysis play in a variety of day-to-day police functions. *Modern Policing Using ArcGIS® Pro* continues Esri's legacy of providing crime analysts the tools and guidance necessary to conduct a spectrum of GIS tasks ranging from making their first crime map to automating complex spatial analysis processes. ArcMap™ software has long provided crime analysts with the tools necessary to support evidence-based police practices. With the transition to ArcGIS Pro, crime analysts now have the opportunity to enhance the efficiency of their workflow and the quality of their crime mapping products.

Interestingly, there has yet to be a singular, compelling work within the crime analysis field designed to convince GIS users to make the switch from ArcMap to ArcGIS Pro or adopt the latter in the first place. Many crime analysts seem hesitant to adopt ArcGIS Pro, despite the substantial leap forward in technological capacity and ease of use. We admit that this could be due to the overwhelming appeal and utility ArcMap has achieved over the years, which in turn delayed our own transition to ArcGIS Pro. However, this same transition led us to realize that crime analysts deserve a work illustrating the powerful tools awaiting discovery in ArcGIS Pro. There also remains an all-too-often ignored audience for a work such as this: the crime analyst's supervisory and command staff. We realize commanders may not have the time to read this work in its entirety but hope crime analysts are able to share this book in some way with command staff to educate them about the ever-increasing value of spatial analysis.

The techniques covered throughout *Modern Policing Using ArcGIS Pro* draw on the authors' experiences working on crime analysis and applied research projects. Eric Piza has over 15 years of experience conducting spatial analysis as a crime analyst at the Newark, New Jersey, Police Department and as an academic researcher at John Jay College of Criminal Justice, the Rutgers Police Institute, and Rutgers Center on Public Security. During this time, he has also taught crime mapping classes on behalf of John Jay College of Criminal Justice, Rutgers University, the International Association of Crime Analysts, and the New York State Division of Criminal Justice Services. Jonas Baughman has cultivated his GIS experience through roles as an analyst, a Real Time Crime Center supervisor, and a strategic analyst in the Office of Chief of Police as a sworn member of the Kansas City, Missouri, Police Department. He has also applied GIS in research settings in the field and teaches crime analysis and GIS techniques through the International Association of Directors of Law Enforcement Standards and Training (IADLEST) and the International Association of Chiefs of Police—Collaborative Reform Initiative Technical Assistance Center (IACP CRI-TAC).

We hope that you can apply the skills covered in this book in your crime analysis work, be it as a police employee, a university researcher, or during your spare time. (Don't worry, we haven't developed any real hobbies, either!)

—Eric L. Piza and Jonas H. Baughman

Acknowledgments

The idea for this book came about soon after I was contacted by Anthony D'Abruzzo, crime analysis training coordinator of the New York State Division of Criminal Justice Services (NYDCJS), about developing ArcGIS® Pro training for crime analysts. Anthony had the foresight to prepare New York State crime analysts for the transition to ArcGIS Pro before much of the field was aware such a transition would even be taking place. I taught this training throughout New York State, with each class providing a sort of laboratory to develop many of the ideas that would eventually become exercises in this book. Anthony deserves a good amount of credit for setting this book in motion. I am also grateful to the crime analysts who attended each seminar for the insight and rigor they brought to the training. After the general book idea was on paper, a number of people at Esri helped bring this idea to life. Mike King heard about our idea for the book and put us in touch with Chris Delaney and John Beck of Esri's Law Enforcement Solutions team. Chris and John helped refine the topic list and provided technical assistance when necessary. Chris and John also originally brokered the larger discussions with Esri Press, which provided guidance and demonstrated patience throughout the process. Thank you to my friend and coauthor, Jonas Baughman, for taking this ride with me. Jonas helped ensure that the exercises in the book offered maximum practical value for crime analysts and their command-level supervisors. Jonas had a much larger role in the success of the book than he gives himself credit for. Thank you to Uma for showing me that creating good work often means navigating some obstacles, and that there is actually joy in the obstacles if you let yourself look for it. Finally, a big thank-you to D.A.P. and everyone at F. Fitus & Associates for constantly reminding me that things are always better when you don't take yourself too seriously.

—Eric L. Piza

I would first like to thank my wonderful wife, Ashleigh, for providing such steadfast, behind-the-scenes support, more than she realizes. She never complained when I had to devote extra time away from family to work on this book. Thank you for believing in me as I worked through what was unfamiliar territory. I love you! My coauthor and friend, Eric Piza, also deserves much gratitude. Eric was patient with me throughout this endeavor, to put it mildly, and he carried more than his fair share of the work on occasion. Writing this book was a stretch for me in many ways, often due to other work-related demands that occurred during 2020. Eric was a true partner from start to finish, and his aptitude and character really shined. I can only hope he and I can work together again in the future. I also wish to thank Esri Press for having confidence in Eric and me to put this book together. I have always wanted to "pay it forward" and share knowledge in some form, including writing a book. I am forever grateful to Esri Press for bringing that dream to fruition.

—Jonas H. Baughman

Introduction

Crime analysis in modern policing

Police are asked to address what at times can seem like an infinite number of problems with a very finite amount of resources. The mismatch between demand for service and available resources to provide such service has only increased over time, as new ideas and priorities have vastly expanded the police agenda. Modern policing includes many functions in addition to crime control, including antiterrorism, procedural justice, third-party policing, and addressing incidents related to suspect mental health (Skogan 2018). To do more with less, police need strong analytic capabilities to diagnose public safety problems, design strategies to control crime and disorder, and evaluate the effect of policies and practices. In this current environment, crime analysts are a prerequisite of effective policing.

The effect of crime analysis on modern policing parallels occurrences from earlier eras (Piza 2019). The early 1900s, which police historians consider the start of American policing's professional era (Miller 1977), saw a great deal of technological innovations, such as the mass production of the automobile and household telephone, that transformed society. Police began to use these new technologies in their day-to-day functions similar to the general citizenry. However, it would be wrong to consider these technologies simply as tools that police incorporated in their mission. Rather, these technologies, in many ways, directly influenced the operational priorities of policing (Kennedy, Caplan, and Piza 2018: 14–15). Automobile patrol replaced foot patrol as the main operational strategy of police. The advent of the 9-1-1 emergency line and two-way radios made the rapid response to citizen calls for service a top priority of American policing. The perceived benefits provided by these technologies, specifically the omnipresence offered by widespread motor vehicle patrol and increased closure of crime offered by rapid response (Wilson 1963), shaped policing through the 1990s and, arguably, remained the cornerstones of most police agencies after the turn of the century (Mastrofski and Willis 2011).

Over time, the perceived benefits of the professional era's primary strategies of omnipresence and rapid response would be rebuked. Declining public confidence in the police combined with an influx of research finding little effect of these strategies led to a reconsideration of the police mission (Weisburd and Braga 2019). Over time, the standard model would become de-emphasized by contemporary police scholars and innovative police managers in favor of strategies collectively known as the focused (Skogan and Frydl 2004) or customized (Sherman 2011) model of policing.

Herman Goldstein (1979) vividly articulated the contrast between such contemporary practices and the standard model of policing in his seminal article "Improving Policing:

A Problem-Oriented Approach." Goldstein (1979) argued that police suffered from a "means over ends syndrome" that placed more emphasis on the organizational structure and operating methods than the substantive outcome of their work. Goldstein argued that this resulted in an incident-based approach, whereby departments respond to individual incidents (many involving the same places and actors) instead of attempt to solve recurring problems that generate these crime incidents in the first place. By the time Goldstein expanded on his ideas in his book *Problem-Oriented Policing* in 1990, a number of additional strategic innovations in policing were being advanced, including hot spots policing (Sherman, Gartin, and Buerger 1989), broken windows (Wilson and Kelling 1982), and situational crime prevention (Clarke and Mayhew 1980). Strategies such as focused deterrence (Kennedy 1997) and third-party policing (Buerger and Mazerolle 1998) were soon after developed with evidence-based policing (Sherman 1998) emerging as a strategic philosophy that encouraged the scientific testing, identification, and use of effective police tactics throughout the field.

The standard model relied on reactive responses to committed crime, but strategies developed after the professional era emphasize proactive police activities and a diversity of approaches for the purpose of preventing crime (Lum, Koper, and Telep 2011; Lum and Koper 2017; Weisburd and Eck 2004). Most if not all of these modern police strategies rely heavily on crime analysis during the design and implementation stages of the intervention. In their seminal text *Crime Analysis for Problem Solvers in 60 Small Steps* (2005), Ron Clarke and John Eck recognized the important role that crime analysts occupy in modern policing. While acknowledging that most crime analysts probably think of their jobs in modest terms (in other words, crunching data for those who do the "real" work), Clarke and Eck (2005) argued that crime analysts possess a unique skillset, which affords them an essential role in addressing public safety problems.

The importance of crime analysis in policing has increased in the years since Clark and Eck's 2005 report. The policing field has experienced a rate of innovation that has outpaced virtually all other government entities, with crime analysis front and center in this strategic evolution (Kennedy et al. 2018). Modern policing strategies require a high degree of analytic precision involving quantitative predictions of where and when crime is most likely to occur (Sherman 2011) as well as theoretical understanding of the causal mechanisms underlying crime patterns (Eck 2017; Sampson, Winship, and Knight 2013) in order to customize strategies to fit local context. Certain strategies, such as hot spots policing or predictive policing, simply cannot occur without the work of crime analysts. Other policing strategies, such as problem-oriented policing, focused deterrence, and CompStat (Santos 2014), also heavily rely on crime analysis. Scholars have further argued that policing would benefit from expanding the role of crime analysts to include functions such as program evaluation (Piza and Feng 2017) and translation of research findings into practice for police managers (Lum and Koper 2017).

Crime and place: The bridge between research and practice

In understanding the ascent of crime analysis in modern policing, it is important to consider a tangential development: the renewed emphasis on the role of place in crime prevention. The work of crime analysts has been facilitated by the increased availability of analytic software, specifically geographic information systems (GIS). The proliferation of GIS and spatial analysis in policing arguably led to a large-scale recognition of place as the optimal unit of analysis for crime prevention efforts (Weisburd 2015). To be clear, police have always focused on place in a certain regard, with police services (for example, patrol) traditionally organized and delivered according to specific units of geography, such as sectors or precincts (Weisburd 2008). However, the problem with measuring problems according to predetermined administrative boundaries is the fact that such boundaries are typically drawn for the convenience of service delivery and may be poor representations of behavioral clusters relating to public safety (Piza, Caplan, and Kennedy 2014a). Therefore, problems residing in shapes and dimensions not reflected in such units of analysis are unlikely to be sufficiently measured through these analytic processes (Sparrow 2016). In that sense, modern policing has reconsidered the notion of place to mean micro-units of analysis, such as block faces or street segments (Weisburd 2008, Weisburd 2015) because these better represent the scale at which public safety problems reside.

The central role of place-based strategies in modern policing shows that crime mapping and spatial analysis are the primary tools of crime analysis. For example, digitized crime data and GIS are key components in the scan and analyze steps of problem-oriented policing (White 2008). Many predictive policing techniques analyze the nature of the urban environment to identify places most at risk of crime (Caplan et al. 2011). Place even plays an important role in offender-based strategies, such as focused deterrence, as the identification of target areas and the analysis of program effects often incorporate GIS methods (Corsaro et al. 2012; Robbins et al. 2017). Sparrow (2011) noted that the evidence-based policing movement is dominated by evaluations of place-based responses to crime. Sparrow (2011) argued that an overemphasis on place may lead researchers to overlook pertinent crime problems not attributable to specific geographies. However, it is important to note that the central role of place in policing is probably due to the observed success of place-based policing strategies. The Committee to Review Research on Police Policy and Practices (Skogan and Frydl 2004) found that geographically focused strategies had the strongest record of effect with a recently published systematic review and meta-analysis similarly supporting the use of hot spots policing (Braga, Turchan, Papachristos, and Hureau 2019). Place-based strategies also offer a more efficient method of policing than offender-based strategies. While places often demonstrate relatively stable crime levels over time, it is well established that individuals experience both short-term and long-term variations in criminal propensity (Agnew 2011). Weisburd (2008, 6), for example, noted that police in Seattle would need to target four times as many people as places to account for 50 percent of the crime incidents between 1989 and 2002.

The crime analysis field was also bolstered by the increased prominence of spatial analysis techniques in the policing literature. Place-based policing research has created a natural bridge between crime analysis and scholarship in a number of ways. For one, many techniques commonly employed by crime analysts, such as hot spot identification and near-repeat crime analysis, are common topics of academic research (Haberman 2017; Hoppe and Gerell 2019). In addition, police analysts are directly contributing to the scientific knowledge base, appearing as authors on a number of recent scientific crime-and-place studies (Haberman and Stiver 2019;

Ratcliffe, Lattanzio, Kikuchi, and Thomas 2019; Telep, Mitchell, and Weisburd 2014). Piza himself is a reflection of this trend, because a number of his research studies were designed or completed while he was a crime analyst at the Newark, New Jersey, Police Department (Piza et al. 2014a, 2014b, 2015; Piza and O'Hara 2014). Crime mapping techniques and the readily accessible nature of crime mapping software have helped to build bridges between crime analysis, academic research, and police practice.

Plan for this book

Crime analysis and GIS have directly influenced the evolution of modern policing. As such, we believe that police agencies lacking robust crime analysis capabilities and the ability to operationalize crime problems to place are at a disadvantage when it comes to the delivery of evidence-based practices. We present this book as our humble attempt to help foster evidence-based policing by empowering crime analysts to continue to contribute high-quality work in support of modern policing. Through ArcGIS® Pro, analysts can conduct problem analyses to help inform the creation of crime prevention strategies, gauge program outputs to measure program implementation, and conduct evaluations to determine whether crime control strategies are having the desired impact.

Modern Policing Using ArcGIS Pro is intended as a guide for all crime analysts, whether GIS novices in their first crime analysis position or seasoned ArcMap™ users transitioning to ArcGIS Pro. Police supervisors and commanders may also find this book helpful, given the range of exercises on strategic analysis, predictive analysis, and workflow automation. The authors have structured the exercises in the book to be accessible and immediately useful and practical to all users. Each chapter includes real-world case studies from crime analysts and applied research projects to demonstrate the spatial analysis techniques necessary to support evidence-based policing. The findings of these case studies are discussed in research and crime analysis highlight sections, with hands-on exercises walking readers through the steps of performing the analysis.

The chapters are structured in a stepwise fashion to incrementally develop the reader's skill level; the book can therefore be worked in sequential order. However, each chapter consists of stand-alone exercises with unique learning objectives, enabling readers to work through the exercises based on a specific task when necessary. We envision that this book will be a constant reference for crime analysts who can consult it as needed during crime analysis projects. The book's format is also beneficial to educators who can cover chapters in the order laid out in the table of contents or in a unique order customized to the precise course curriculum and expertise level of the students (undergraduate students, graduate students, working crime analysts, and so on). In essence, *Modern Policing Using ArcGIS Pro* seeks to become the quintessential tool to help crime analysts and commanders bring value to their agencies' operations through crime and data analysis.

Chapter 1
Exploring ArcGIS Pro

Overview

This chapter provides a foundational understanding of ArcGIS® Pro software. You will learn about the basic layout and functionality of ArcGIS Pro, including how to navigate the software to create GIS documents, import data, and access the ribbons and panes.

You will acquire the following skills upon completing the exercises in this chapter:
- Creating a new ArcGIS Pro project
- Accessing ArcGIS Online and data portals
- Importing data and creating folder connections
- Navigating the ribbon and project panes
- Exploring the map view
- Symbolizing layers

Exercise 1a: Connect to ArcGIS Online and data portals

Crime analysis highlight 1a: Using portals for agencywide analysis

Once you conduct a variety of analyses and generate maps, what do you do with the final products? Who would benefit from seeing this information? Police departments often have large amounts of data and information around various programs they implement, with no easy way to present it to the communities they serve. The Rochester (New York) Police Department facilitates agencywide analysis by using ArcGIS Online and data portals. By housing data in our online portal, we can share information on the various partner efforts, as well as data on the crimes in the city of Rochester and the identified hot spots. This data is presented in a live and interactive ArcGIS dashboard, in which users can select specific crime types, hot spots, or dates to see what the crimes look like for their selection. We also use ArcGIS Online to create ArcGIS StoryMaps℠ stories, which allow users to present data, maps, images, and qualitative information in one clear, user-friendly story. Dashboards and stories allow all our local agency partners to be transparent with the community and allow community members to learn more about the areas they live in and the work that public safety is doing.

—Kayla Macano, director of operations, Monroe County (New York) Gun Involved Violence Elimination (GIVE) initiative

ArcGIS Pro represents a significant advancement in ArcGIS. A distinguishing feature of ArcGIS Pro is its integration with ArcGIS Online. The accessibility to ArcGIS Online allows ArcGIS Pro users to share resources within and among organizations and Esri® user communities. ArcGIS Online also provides access to a wealth of data and tools created and shared by the Esri community. All exercises in this book are performed using data from ArcGIS Online.

Create a working folder

You will create a working folder on your computer to house all data and exercise outputs for this book before obtaining data for this chapter. All the book exercises will work off a single folder that you will create on drive C. Keeping a uniform folder will ensure easy access to and organization of all exercise-related files.

1   Browse to your computer's C drive or desktop, right-click New, and click Folder.

2   Name the new folder **ModernPolicing.**

You will now create an ArcGIS Pro project to complete the chapter 1 exercises.

3   Start ArcGIS Pro.

4   Sign in using your ArcGIS Online organizational account.

5   Under the New heading, click Map to create a map project.

6   In the Create A New Project window, type **Chapter1** in the Name field.

7   To the right of the Location field, click the folder and browse to C:\ModernPolicing.

ArcGIS Pro creates a project and encompassing folder, both named Chapter1.

8 In the Create A New Project window, click OK.

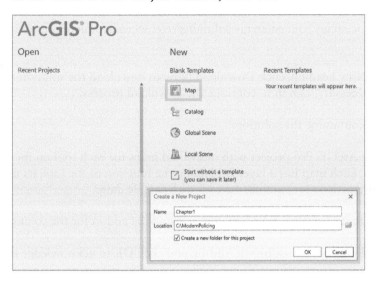

By default, ArcGIS Pro creates several complementary files along with each new project (.aprx). The newly created Chapter1 folder includes a geodatabase (.gdb), in which all newly created data is stored by default; a toolbox (.tbx), in which users can customize geoprocessing functions and models; an Index folder, which serves as a searchable index of all items added to a project; and an ImportLog folder, which houses XML files associated with the project items.

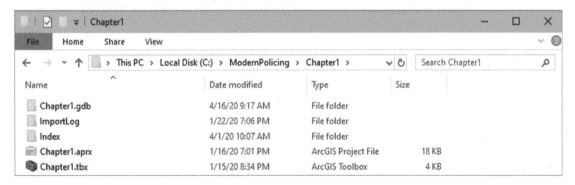

Download and install the ArcGIS Crime Analysis Solution

ArcGIS Solutions provides collections of preconfigured maps, apps, tools, and workflows based on the needs of specific industries. The Crime Analysis Solution is a configuration of ArcGIS Pro that can be used by crime analysts to perform a series of analysis functions. The Crime Analysis tab provides easy access to functionality related to crime analysis. The entire solution can be deployed to your ArcGIS Online organization, allowing members of your organization to use it. It can also be installed on your desktop computer. For the exercises in this book, you will download and install the solution to your desktop computer.

1 Exit ArcGIS Pro, if necessary.

2 Browse to https://doc.arcgis.com/en/arcgis-solutions/reference/introduction-to-crime-analysis.htm.

3 Under the Deploy Now heading, click Download Now to download the solution to your computer. If necessary, sign in to continue the download process.

4 Extract the zip file containing the solution.

   The file contains an ArcGIS Pro project with individual maps for each tool on the Crime Analysis tab. Each map has a layer explaining the function of the tool, its use to analysts, and instructions for running the tool with sample data.

5 Double-click the CrimeAnalysis.esriAddinX file to install the add-in for the solution.

6 In the installation dialog box, click Install Add-In, and click OK to acknowledge the installation.

7 Start ArcGIS Pro.

8 On the start-up screen, click Settings, and then click Add-In Manager.

9 Under Add-Ins, confirm that Crime Analysis is listed.

   You can manage the visibility of the Crime Analysis tab from the Project menu.

10 Click Options.

11 From the Options menu, select Crime Analysis. Confirm that the Show Crime Analysis Tab option is checked.

   You are now ready to use the ArcGIS Crime Analysis Solution for the remainder of the book.

Download and install the data

With the new ArcGIS Pro project created within the home folder, you will now obtain the data necessary for the chapter 1 exercises.

1 Go to www.arcgis.com and sign in with your ArcGIS Online account credentials.

Always be mindful that ArcGIS Online contains a wealth of publicly available data that you can import into your ArcGIS Pro projects. As an example, searching for the term *crime analysis* results in thousands of individual sources of content and dozens of groups actively sharing data. From our experience, we've found useful data for our jurisdictions that complemented the data collected by our agencies. You will be interacting with crime data from ArcGIS Online throughout this book.

You will download the chapter 1 data for the following exercises.

2 Type **Modern Policing Using ArcGIS Pro (Esri Press)** in the search box and click the Groups tab. Make sure that the Only Search In Your Organization option is turned off.

3 Click the group Modern Policing Using ArcGIS Pro (Esri Press).

4 On the group heading, click Content.

5 Click Layers for Chapter 1: Modern Policing Using ArcGIS Pro and download it.

6 Locate the zip file you downloaded to your local drive, right-click, and extract it to C:\ModernPolicing\Chapter1.

You have now created a subfolder named layers_chapter1 in the Chapter1 folder. This subfolder contains the data you will use for the remainder of this chapter's exercises.

la

Exercise 1b: Explore the ribbon and project panes

This section introduces the basic interface and functionality of ArcGIS Pro. Before proceeding through the exercise steps, make sure the Chapter1.aprx project is open and visible on your monitor.

The main ArcGIS Pro window is divided into four primary components. Each component can be customized in terms of size, location in the project, and ability to minimize panes or stack them on top of one another. You can change the positioning of a pane by clicking the header and dragging the pane to another area in the project.

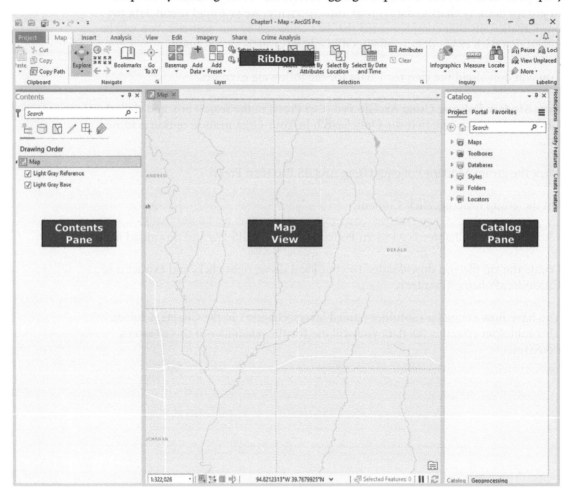

The ribbon sits horizontally on top of the project. The ribbon organizes ArcGIS Pro functionality into a series of tabs. Core tabs are those that are always open, whereas contextual tabs appear only when the application is in a particular state (for example, contextual table tabs appear when a table is open). The Map tab is visible by default when an ArcGIS Pro project opens. As discussed in the next exercise, the Map tab contains the main navigation tools that you will use to interact with your map.

Change the basemap

1 On the Map tab, in the Layer group, click Basemap.

2 From the Basemap drop-down list, click Light Gray Canvas.

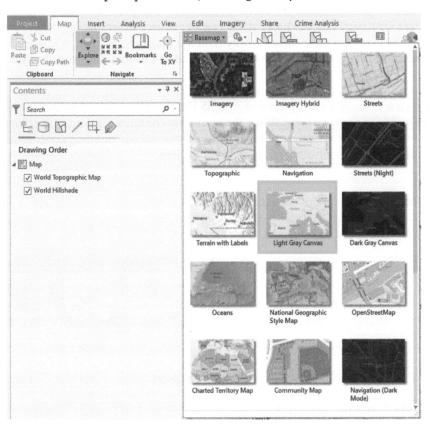

Each tab on the ribbon contains sets of tools that you will use for various purposes. The Crime Analysis toolbar is of particular interest to crime analysts. It contains several new and out-of-the-box tools meant to streamline the crime analysis process. It will be accessed often throughout the exercises in this book.

Note: If you do not see the Crime Analysis tab, the solution is not activated on your computer. To activate the Crime Analysis tab, click the settings icon on the ArcGIS Pro start screen. Click Options, and click Crime Analysis. In the Display Settings window, make sure that the Show Crime Analysis Tab is checked.

The Contents pane provides a list of contents in the active view, such as layers and tables contained in a map.

The Contents pane opens in the List By Drawing Order view by default. The view can be changed to one of six others: List By Data Source, List By Selection, List By Editing, List By Snapping, List By Labeling, or List By Perspective Imagery. The most appropriate view varies on the basis of the tasks you are performing in ArcGIS Pro. Currently, the Contents pane includes only the base imagery and reference labels for the basemap. The Contents pane becomes populated as items are added to the project.

The map view is the main work area of ArcGIS Pro. Layers activated in the Contents pane are visible in the map view. When the map view contains 3D data, it is referred to as a *scene*. The view automatically updates to show the surrounding geography of the data layers when you add spatial data to your map.

Add data to the map view

1   On the Map tab, in the Layer group, click Add Data.

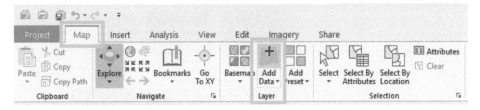

2   Browse to C:\ModernPolicing\Chapter1\layers_chapter1.

3   Click KCPD_Districts_Divisions.lpkx.

4   Click OK.

The map view now shows the districts and divisions of the Kansas City, Missouri, Police Department. The layer is now included in the Contents pane.

The Catalog pane allows you to access in one place all the items associated with the current project. The Catalog pane provides easy access to all the content in your project's encompassing folder (in this case, Chapter1) via the Folders heading. As will be demonstrated in chapter 3, the Contents pane is commonly your first stop when creating content in your project.

You can access data outside the active project folder by establishing a direct connection to the appropriate folder using the Add Folder Connection tool. That will allow single-click access to the folders of interest on your computer or network.

Create a folder connection

1   In the Catalog pane, right-click Folders and click Add Folder Connection.

2   In the Add Folder Connection window, browse to C:\ModernPolicing\Chapter1, click the layers_chapter1 folder, and click OK.

3   In the Catalog pane, expand Folders to display the contents.

The folder connection for layers_chapter1 now appears in the Catalog pane.

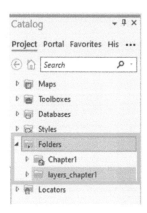

Expanding the layers_chapter1 heading will reveal all datasets available for use in your ArcGIS Pro project. Layers contained in connected folders are listed, even if they have been added to the Contents pane. In the layers_chapter1 folder, KCPD_Districts_Divisions.lpkx, which was added previously, is listed with other shapefiles that are currently not displayed in the project. In addition to using the Add Data button, you can add layers to the Contents pane by dragging them onto the map from the Catalog pane.

Add data from the Catalog pane

1   In the Catalog pane, in the layers_chapter1 folder connection, select Homicides_KC_Qtr1, Homicides_KC_Qtr2, Homicides_KC_Qtr3, and Homicides_KC_Qtr4. Select multiple layers by pressing the Ctrl key simultaneously.

2   Drag your pointer over the map view while clicking the left mouse button simultaneously.

3 Let go of the left mouse button to add these layers to the map.

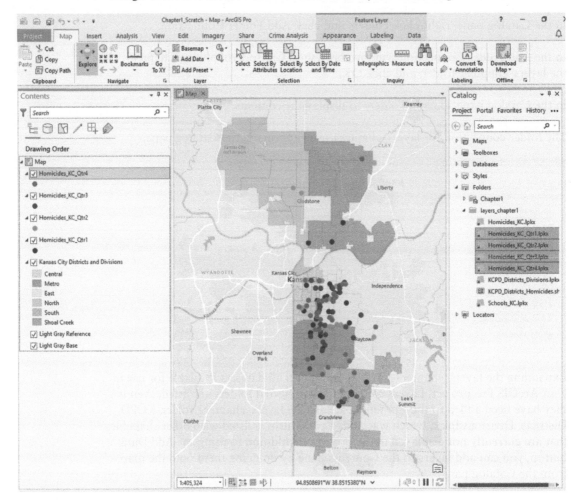

In addition to the folders on your computer, you have the ability to access data portals to find content for your project.

4 Click the Portal tab in the Catalog pane.

The first section of the Portal pane shows data that the user currently signed in to ArcGIS Pro has published to ArcGIS Online. The second displays content that has been added to favorites. The third section displays all the ArcGIS Online user groups to which the user belongs. Double-clicking a group name displays all the content shared by members of the group. The fourth section allows users to see content from the ArcGIS organization to which they belong. The fifth section shows content from ArcGIS Online. The sixth section provides access to ArcGIS Living Atlas of the World, a collection of global geographic information curated by Esri with contributions from its partner and user communities.

> **Note:** For more information about ArcGIS Living Atlas, see livingatlas.arcgis.com.

Exercise 1c: Use ArcGIS Pro navigation tools

After populating your map with data, you will interact with your map by zooming in and out, creating bookmarks of places of interest, and exploring the underlying attributes of your spatial data. ArcGIS Pro offers enhanced navigation tools to facilitate this process.

All navigation tools are accessible on the Map tab, in the Navigate group. The Explore tool allows you to zoom in and out, pan, and select features. The Explore tool combines the functionality of tools that were separate in earlier versions of Esri software into a single tool. You will now use the Explore tool to navigate the map.

Use the Explore tool

1   On the Map tab, in the Navigate group, click the Explore tool.

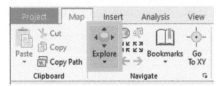

The Explore tool allows you to change the scale of your map view. Scrolling the mouse wheel up and down zooms out and in on your map, respectively. Holding the left mouse button allows you to pan around the map. Holding the Shift key allows you to draw a rectangle around the area to which you want to zoom. You will typically use a variety of these functions when navigating within ArcGIS Pro.

2   Zoom to the Metro patrol division of Kansas City.

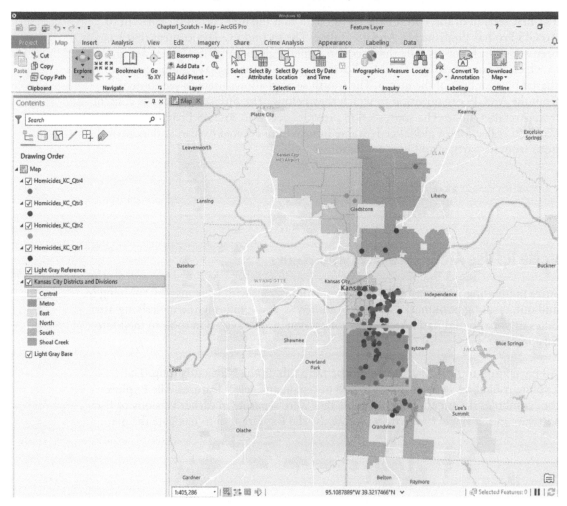

A number of homicides have been reported in the Metro patrol division. You will now use the Explore tool to learn more about these events.

3 Click the easternmost third-quarter homicide (purple dot) in the Metro division.

The pop-up window shows the attributes associated with this point.

You will now create a bookmark for the Metro patrol division to allow you to quickly return to this view later.

Create a bookmark

1 On the Map tab, click Bookmarks, and click New Bookmark.

2 In the Create Bookmark window in the Name field, type **Metro Patrol Division.**

3 In the Description field, type **This bookmark displays the Metro Patrol Division of the Kansas City Police Department.**

4 Click OK.

This newly created bookmark now appears on the Bookmarks drop-down list. Click the Metro Patrol Division button whenever you want to automatically zoom to the same area.

> **Note:** You can update the spatial extent, name, and description of bookmarks with the Manage Bookmarks option of the Bookmarks tool.

Use the zoom and extent tools

Now that you are done working in the Metro patrol division, you will zoom out so that the Kansas City police divisions can be viewed in their entirety. ArcGIS Pro provides several navigation tools you can use in this instance.

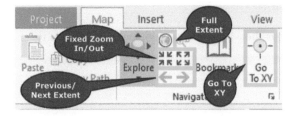

The Fixed Zoom In and Fixed Zoom Out tools allow for quick zooming in to and out from a fixed distance. These tools are not as gradual of a zoom compared with using your mouse, but they provide a quick way to zoom in or out for a brief look at your data.

The Previous Extent and Next Extent tools work the same way as the back and forward buttons in an internet browser. If you zoom in on your map, view your data, and then want to return to the previous zoom, click the Previous Extent tool (the left blue arrow). Click the Next Extent tool (the right blue arrow) to move back to the extent just used.

The Full Extent tool provides a way to instantly zoom out from your map to the maximum extent represented by your data. **Warning:** Clicking this tool can produce undesired results (easily reversible with the Previous Extent tool), depending on your data. For example, if you are using a basemap from the Esri portal, it often contains worldwide information. Clicking the Full Extent tool zooms out from your map to this worldwide scale.

The Go To XY tool opens a small, separate toolbar. You can enter specific coordinate information to zoom to that location on your map.

You can also change the extent of your map by relating the scale to a layer in the Contents pane. You will now do that to return to the citywide extent.

1   In the Contents pane, right-click Kansas City Districts and Divisions, and click Zoom To Layer.

Your map view is now back to the citywide extent.

Exercise 1d: Symbolize features (part 1)

In this exercise, you will symbolize the point and polygon data in your map. You will begin by organizing the four homicide layers into a single group layer.

1 In the Contents pane, select each of the four homicide layers (in other words, quarter 1, quarter 2, quarter 3, and quarter 4).

2 Right-click any of the selected layers and click Group.

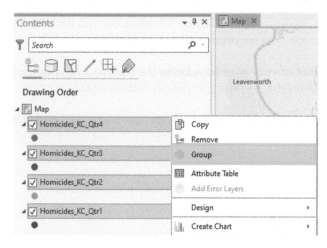

> **Note:** All the layers must be selected while clicking Group. Any layers that are not selected will not be added to the Group layer.

A new group named New Group Layer appears in the Contents pane. You will now give the layer a new name.

3 Double-click New Group Layer to open the Layer Properties window.

4 In the Layer Properties window, type **Homicide: 1 Year** in the Name field.

5 Click OK.

> **Note:** You can ungroup layers the way that you group them: right-click the layer and click Ungroup.

6 In the Contents pane, uncheck the boxes next to the layer names to turn off the visibility for all layers in the group except Homicides_KC_Qtr4.

A contextual Feature Layer tab appears on the ribbon when a layer is selected.

7 On the Feature Layer contextual tab, click the Appearance tab.

8 Click the Symbology button.

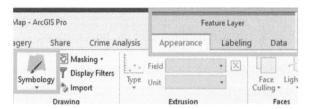

9 In the Symbology pane, click the shape to the right of the symbol heading.

The default symbology is Single Symbol, which indicates that all features in a layer are identical (for example, all are points). You can change the appearance of all points in your layer by clicking the symbol to activate the Format Point Symbol pane.

10 In the Format Point Symbol pane, click the Circle 3 symbol.

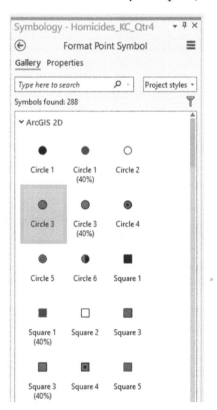

All the points in Homicides_KC_Qtr4 are now symbolized using the Circle 3 style.

Symbolize by unique values

Using a single symbol makes sense in certain instances, but this is not always the case. For example, all the homicide layers contain an attribute that captures the nature of the crime (shooting, stabbing, and so on). It may therefore be more useful to apply unique symbols for each homicide type.

1   In the Symbology pane, click the back arrow near the top of the tool window.

2   Click the arrow for Primary Symbology and click Unique Values.

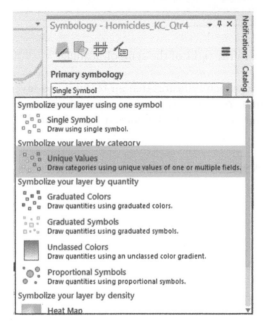

3   In the Field 1 drop-down list, click Cause_of_d to symbolize based on the cause of death.

ArcGIS Pro automatically fills in the table on the bottom half of the Symbology tool, outlining the types of offense categories available in the dataset.

You will now symbolize each of the offense categories using a unique color.

4   Click the symbol to the left of Other.

The Format Point Symbol window within the Symbology tool appears.

5 Click Circle 3, and then click the Properties tab.

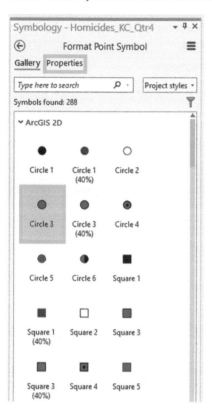

6 Click the arrow to the right of Color, and click Gray 70%.

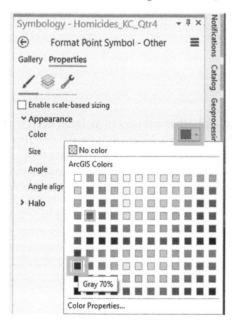

7 Click Apply.

8 Repeat steps 1 through 8 for the Shooting and Stabbing points. Apply red circles to Shooting points and blue circles to Stabbing points.

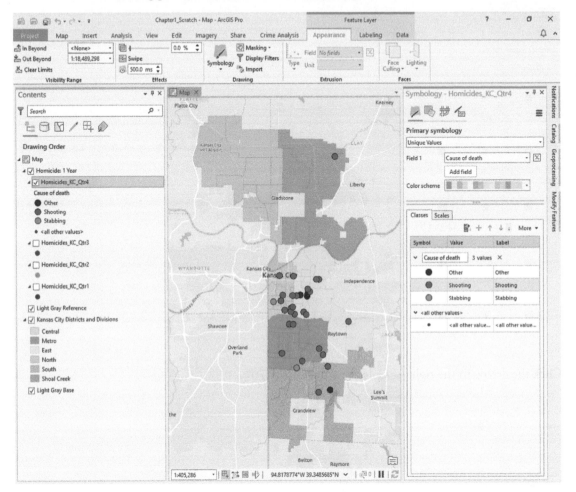

The Symbology pane lists a category named <all other values> by default, even if all the attribute values are currently displayed on the map. You will now remove this category from the map.

9 Under Primary Symbology, on the Classes tab, click the More button located above the individual symbols for your point data.

10 Uncheck the Show All Other Values box.

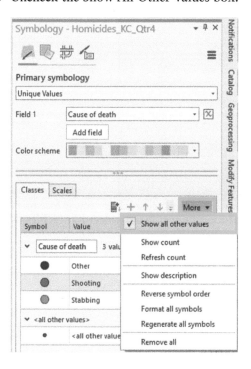

> **Note:** Sometimes it may not be necessary to display all categories of an attribute on your map. To remove a category, right-click it in the Symbology pane and click Remove.

Import symbology

With the quarter 4 homicides symbolized, you will now symbolize the homicide points of the other three quarters. You can manually symbolize these layers by following the preceding steps. However, considering that all the homicide layers have the same attributes and attribute categories, you can quickly apply the symbology of Homicides_KC_Qtr4 to the other layers.

1 In the Contents pane, click Homicides_KC_Qtr3 and turn on the layer.

2 On the Appearance tab, click the Symbology button.

3 In the Symbology pane, click Import Symbology.

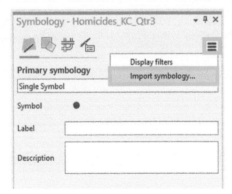

4 In the Apply Symbology From Layer geoprocessing tool pane, select Homicides_KC_Qtr4 from the Symbology Layer drop-down list.

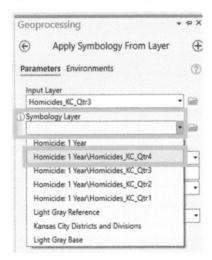

5 Click Run.

The Homicides_KC_Qtr3 points now have the identical symbology style as Homicides_KC_Qtr4.

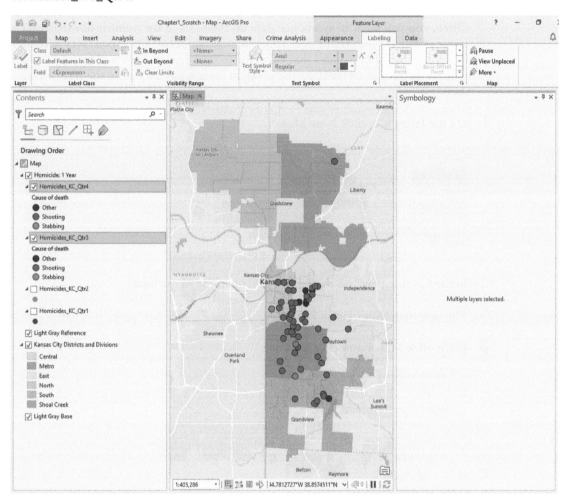

6   Repeat steps 1 through 5 to import this symbology to Homicides_KC_Qtr1 and Homicides_KC_Qtr2.

7   Turn the layers on and off to get a sense of where homicides occurred in each quarter.

8   When you have finished, leave the Homicides_KC_Qtr4 layer on and turn off the other layers in the group.

Exercise 1e: Symbolize features (part 2)

You symbolized data on the basis of the feature attributes in the prior exercise. Often, you will want to symbolize data on the basis of the number of events, such as crime, that have occurred there. You will now symbolize a polygon layer according to the number of homicides that occurred in each feature. You will first add a new map to the project to visualize the polygon layer separately from the rest of your data.

1   On the Insert tab, click New Map.

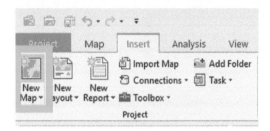

2   In the Catalog pane, on the Project tab, open the Maps folder.

The new map (Map 1) is added.

3   Right-click Map 1 and click Rename.

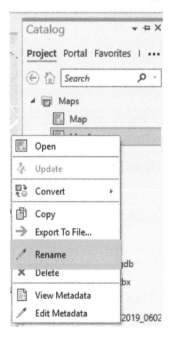

4   In the search box, type **Graduated Colors**.

5 In the Catalog pane, under Folders, open the layers_chapter1 folder, and click KCPD_Districts_Homicides.

6 Drag KCPD_Districts_Homicides to the Graduated Colors map.

> **Note:** You can also add the layer to the map by right-clicking and selecting Add To Current Map.

7 In the Contents pane, right-click KCPD_Districts_Homicides, and click Attribute Table.

To the right of the attribute table, the Homicides field lists the number of homicide incidents that occurred in each division. You will now symbolize each division on the basis of its crime count.

8 Close the attribute table.

Symbolize by graduated colors

1 In the Contents pane, click KCPD_Districts_Homicides.

2 On the Appearance tab, in the Drawing group, click the Symbology down arrow, and choose Graduated Colors.

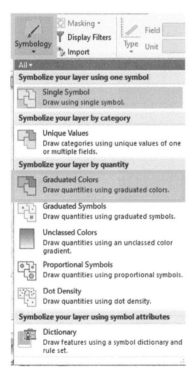

3   In the Symbology pane, from the Field drop-down list, click Homicides.

The KCPD_Districts_Homicides layer is now symbolized according to the number of homicides that occurred in each district.

> **Note:** ArcGIS Pro assigns the natural breaks method to classify values by default. Six additional classification schemes are available from the Method drop-down list in the Graduated Colors pane. ArcGIS Pro provides definitions of each classification scheme in the list.

Although the graduated colors scheme provides the intended visual for the KCPD_Districts_Homicides layer, the opaque symbology may not be appropriate in all instances. For example, it may be helpful to view the underlying basemap while the polygon layer is activated so that you can identify the names of streets within the high-crime divisions. You will add transparency to the layer to accomplish this.

4   In the Contents pane, click KCPD_Districts_Homicides.

5   On the Appearance tab, in the Effects group, set the transparency level to **50%**.

The underlying basemap is now visible beneath the polygon layer.

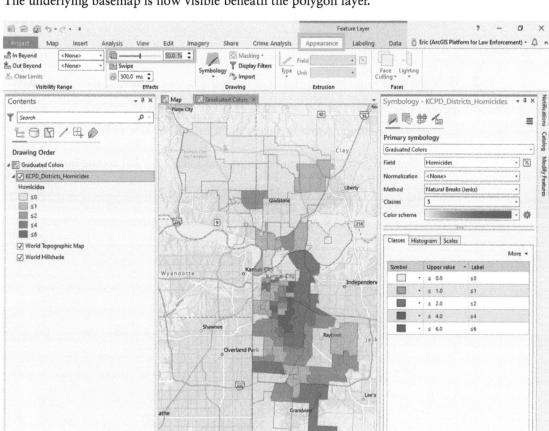

Exercise 1f: Link map views

In the prior exercises, you created two maps displaying different types of crime and police data in Kansas City, Missouri. It may be helpful to view and interact with the maps simultaneously in certain circumstances. You will use the Link Views function. When linking is enabled, all linked map views automatically update to the currently active view.

Prior to linking your maps, you will reposition the map views so that they appear side by side.

1   With both the Map and the Graduated Colors maps open, click the Graduated Colors map tab.

1f

2   Hold the left mouse button and drag the tab downward toward the map view.

The docking icon appears as you drag your mouse. The docking icon is used to arrange panes within a window in a stacked or side-by-side fashion.

Hover over the dock display on the right, and release the left mouse button.

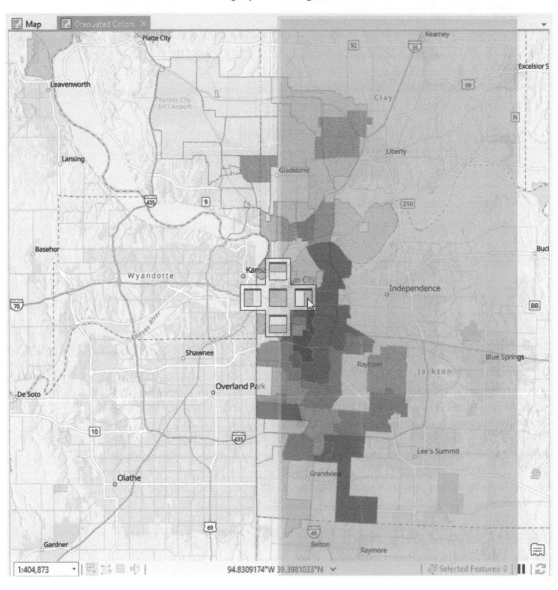

The Map and the Graduated Colors maps now appear side by side.

3 Click the Map tab to make it the primary view.

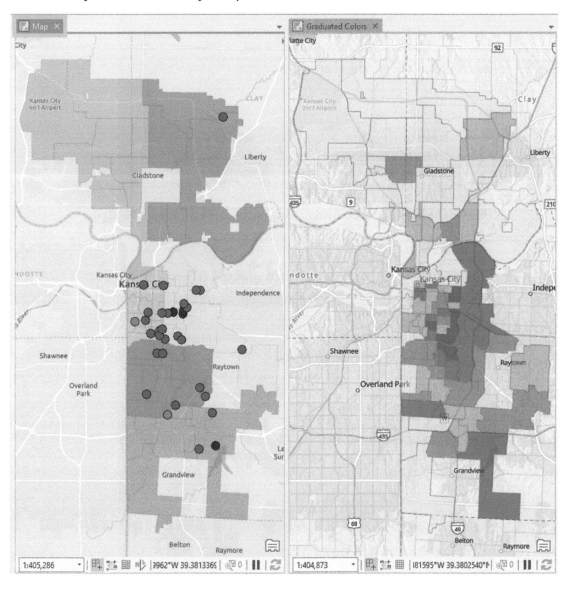

1f

4 On the View tab, click the Link Views down arrow, and select Center And Scale.

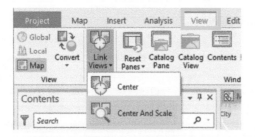

> **Note:** Linking map views using the Center option updates only the center of the linked map, leaving the scale unchanged.

5 On the Map tab, click Bookmarks, and click Metro Patrol Division.

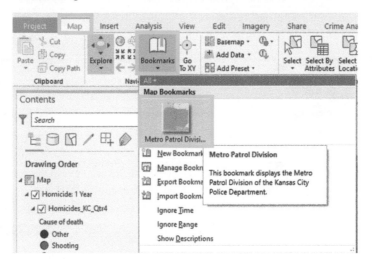

The two maps simultaneously zoom to the Metro Patrol Division.

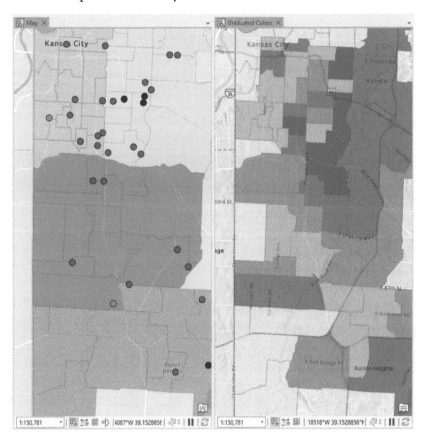

Updating the scale of the map using the Explore tool similarly updates the view of both maps at once.

6   On the Map tab, in the Navigate group, click the Explore Tool.

7 Holding the Shift key and the left mouse button, drag your mouse pointer to draw a rectangle around the two northernmost homicides in the Metro division.

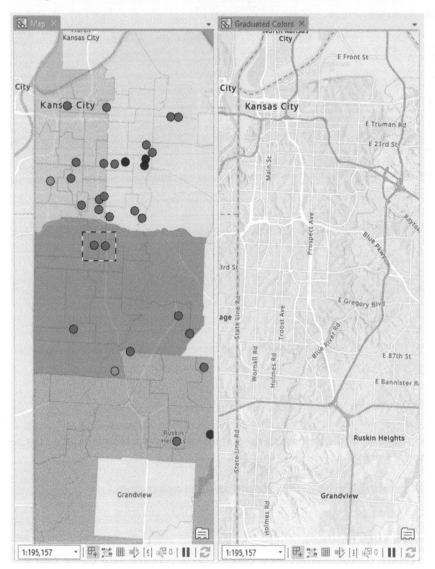

Both map views are now focused on the two homicide points.

Summary

This chapter introduced you to ArcGIS Pro, including its core layout, functionality, and processes for symbolizing data. The chapter covered important concepts that GIS users need to build a foundation, including creating an ArcGIS Pro project, navigating the ribbon format, and connecting to and importing data. New ArcGIS Pro users should be comfortable with the material from this chapter, which serves as the cornerstone for the remainder of the book.

Chapter 2
Geoprocessing and selecting data

Overview

This chapter introduces you to the ArcGIS Pro geoprocessing tools and demonstrates how to select features of interest in data layers. You will learn how to browse toolboxes, search for specific geoprocessing tools, and run common geoprocessing functions used to identify features of interest within data layers.

You will acquire the following skills upon completing the exercises in this chapter:
- Accessing the Geoprocessing pane
- Creating individual and dissolved buffers
- Clipping underlying features using another layer
- Merging separate features into a single layer
- Conducting attribute and location selections
- Selecting features by date and time

Geoprocessing refers to any GIS operation that manipulates data. Typically, a geoprocessing operation takes an input dataset, performs an operation on that dataset, and returns the results as an output dataset. Geoprocessing tools are used to perform geoprocessing functions. Geoprocessing tools are stored in toolboxes and run from the Geoprocessing pane. There are three types of geoprocessing tools: built-in tools, model tools, and script tools. This chapter focuses on built-in tools.

> **Note:** Model tools and script tools are commonly used to automate geocoding processes either through ModelBuilder™ or from a Python script. These types of geoprocessing tools are covered in chapter 8 ("Automating crime analysis processes").

Download and install the data

Before working on the exercises, you will obtain the necessary data.

1 Go to www.arcgis.com and sign in with your ArcGIS Online account credentials.

2 Type **Modern Policing Using ArcGIS Pro (Esri Press)** in the search box, and click the Groups tab. Make sure that the Only Search In Your Organization option is turned off.

3   Click the Modern Policing Using ArcGIS Pro (Esri Press) group.

4   On the group heading, click Content.

5   Click the chapter 2 file and download it.

6   Locate the zip file you downloaded to your local drive, right-click, and extract it to C:\ModernPolicing.

This will create a folder named Chapter2 in the ModernPolicing folder.

This folder contains an ArcGIS Pro project and data you will use for the exercises in this chapter.

Exercise 2a: Use the Buffer tool

Crime analysis highlight 2a: Using buffers to help coordinate reentry services

Working with other agencies is a key part of law enforcement as a whole and crime analysis in particular. As such, I spend time every week working with my agency's law enforcement partners. I recently worked with a local state probation and parole officer who monitors sex offenders in our area to help her map out where her clients were prohibited from living. In Missouri, certain sex offenders cannot live within 1,000 feet of a school or day-care facility. Probation and parole officers have to keep this restriction in mind when supervising a sex offender's transition back into the community. To aid in this process, I created a map displaying all the schools and day cares in our city surrounded by a 1,000-foot buffer. The probation and parole officer uses these maps when she meets with her clients to plan living arrangements. These maps assist her clients in selecting housing that both complies with the law and is placed within close proximity to necessary social services and places of employment.

—Kyle J. Stoker, crime analyst, Raytown (Missouri) Police Department

Create buffer features

1   Open Chapter2.aprx in C:\ModernPolicing\Chapter2.

The project shows a map named Buffer & Clip. This map displays all public school locations in Kansas City, Missouri. You will work with this data to generate buffers around each of the school points.

> **Note:** The public schools feature layer is saved in Chapter2.gdb (the geodatabase) as Public_
> Schools_KC, which is different from how it appears in the Contents pane: Public Schools
> (KC). For ease of interpretation, the layer's name was altered on the General tab of the Layer
> Properties window, which can be opened by double-clicking the layer in the Contents pane.
> While this changes how the name appears in the Contents pane, the name of the feature layer is
> unchanged. To change the name of the feature layer, right-click Public_Schools_KC in Chapter2.
> gdb and select Rename. This applies to all feature layers used throughout the book.

2 On the Analysis tab, in the Geoprocessing group, click Tools.

The Geoprocessing pane appears to the right of the project. The text box at the top of the Geoprocessing pane allows you to search for tools.

3 In the search text box, type **Buffer** and press Enter.

> **Note:** ArcGIS Pro lists frequently used tools in the Favorites section as you work with
> geoprocessing tools. This allows you to run the tool without first having to open the search pane.

4 In the search results, click Buffer (Analysis Tools).

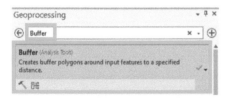

5 In the Buffer pane, select Public Schools (KC) from the Input Features list.

6 In the Output Feature Class text box, type **Buffer_Schools**.

> **Note:** Geoprocessing outputs are saved in the current project geodatabase (in this example,
> Chapter2.gdb) by default. The folder to the right of the text box enables you to change the saved
> location of the output feature class.

7 In the Distance text box, type **1000** and select Feet from the list.

8 Leave all other options as is.

9 Click Run.

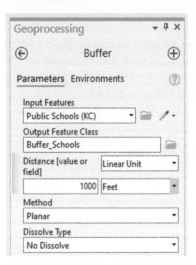

The newly created Buffer_Schools layer now appears in the Contents pane. You will inspect the new layer by zooming in closer and viewing the attribute table.

10 On the Map tab, in the Navigate group, click Bookmarks, and click Buffer Zoom.

11 In the Contents pane, right-click Buffer_Schools, and click Attribute Table.

Here you see that ArcGIS Pro generated 165 individual buffers, one for each of the schools in the original feature class. Each buffer is an individual feature in the layer, even when it overlaps with other buffers.

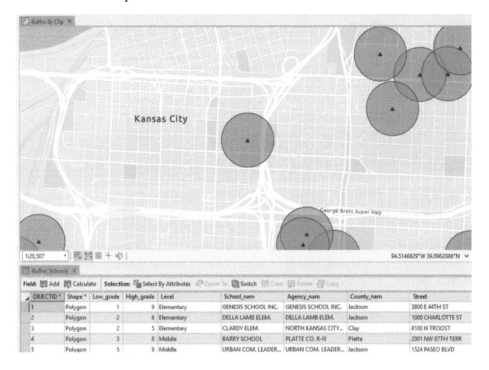

OBJECTID *	Shape *	Low_grade	High_grade	Level	School_nam	Agency_nam	County_nam	Street
1	Polygon	1	9	Elementary	GENESIS SCHOOL INC.	GENESIS SCHOOL INC.	Jackson	3800 E 44TH ST
2	Polygon	2	6	Elementary	DELLA LAMB ELEM.	DELLA LAMB ELEM.	Jackson	1000 CHARLOTTE ST
3	Polygon	2	5	Elementary	CLARDY ELEM.	NORTH KANSAS CITY...	Clay	8100 N TROOST
4	Polygon	3	8	Middle	BARRY SCHOOL	PLATTE CO. R-III	Platte	2001 NW 87TH TERR
5	Polygon	5	9	Middle	URBAN COM. LEADER...	URBAN COM. LEADER...	Jackson	1524 PASEO BLVD

Dissolve buffers into one

You can use the dissolve option to create a single buffer for all schools rather than a separate feature for each individual school.

1 Open the Buffer tool.

2 In the Geoprocessing pane, select Public Schools (KC) from the Input Features list.

3 In the Output Feature Class text box, type **Buffer_Schools_Dissolve**.

4 In the Distance text box, type **1000** and select Feet from the list.

5 From the Dissolve Type list, select Dissolve All Output Features Into A Single Feature.

6 Click Run.

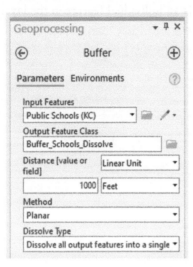

The new Buffer_Schools_Dissolve layer now appears in the Contents pane. Rather than 165 separate features, this file consists of only a single feature with overlapping buffers merged. This is reflected in both the map and the attribute table.

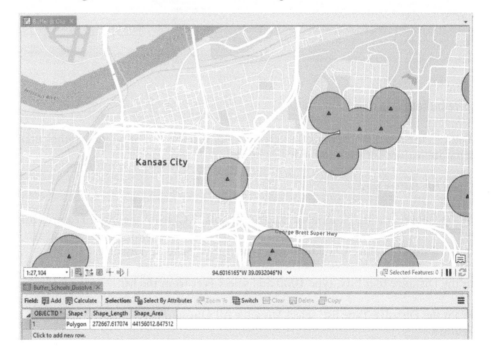

Dissolve buffers by attribute

The Buffer tool can also dissolve buffers according to categories contained in the attribute table. Using this option creates a dissolved buffer for each unique attribute value. You will create a buffer file dissolved by level of school (for example, elementary school, high school, and so on).

1 Open the Buffer tool.

2 In the Geoprocessing pane, select Public Schools (KC) from the Input Features list.

3 In the Output Feature Class text box, type **Buffer_Schools_Level**.

4 In the Distance text box, type **1000** and select Feet from the list.

5 From the Dissolve Type list, select Dissolve Features Using The Listed Fields' Unique Values Or Combination Of Values.

6 From the Dissolve Field(s) list, select Level.

7 Click Run.

The new Public_Schools_KC_Level appears in the Contents pane. This buffer file consists of 11 features, one for each school level in the city.

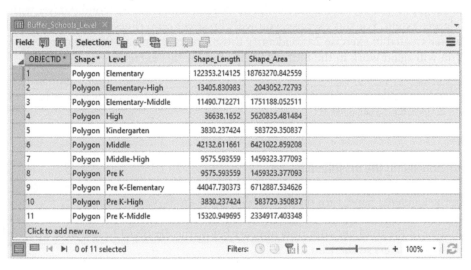

OBJECTID *	Shape *	Level	Shape_Length	Shape_Area
1	Polygon	Elementary	122353.214125	18763270.842559
2	Polygon	Elementary-High	13405.830983	2043052.72793
3	Polygon	Elementary-Middle	11490.712271	1751188.052511
4	Polygon	High	36638.1652	5620835.481484
5	Polygon	Kindergarten	3830.237424	583729.350837
6	Polygon	Middle	42132.611661	6421022.859208
7	Polygon	Middle-High	9575.593559	1459323.377093
8	Polygon	Pre K	9575.593559	1459323.377093
9	Polygon	Pre K-Elementary	44047.730373	6712887.534626
10	Polygon	Pre K-High	3830.237424	583729.350837
11	Polygon	Pre K-Middle	15320.949695	2334917.403348

Click to add new row.

0 of 11 selected   Filters: — + 100%

Exercise 2b: Use the Clip (Analysis) tool

Crime analysis highlight 2b: Using the Clip tool to focus crime analyses on areas of interest

As an analyst working in city government, you need to provide focused maps that policy makers can read and understand in an instant. This is made increasingly possible because cities are continually expanding the scope and accuracy of their GIS data. This gives the analyst a great amount of flexibility to create customizable maps of even the most specific crime trends for operational management and policy makers. For example, in response to an increase in crime incidents on a popular trail that ran along a river, we removed the crime incidents that were not within the trail area boundaries using the Clip tool. This allowed us to visually inspect the trail area, and the distribution of crime along it, to quickly identify problem areas that we then shared with our partners across city agencies. Without the Clip tool, it would be very easy to become distracted by crime incidents outside the area of interest or other features included in the map that are only for context.

—David Hatten, former GIS analyst, City of Philadelphia (Pennsylvania); doctoral candidate, John Jay College of Criminal Justice/City University of New York Graduate Center

The Clip tool, much like the Buffer tool, allows you to quickly create a dataset on the basis of a specific need. Analysts may often find themselves with a dataset that is larger than needed or that covers a geographic extent that far exceeds other data on the map. The Clip tool can be used in situations in which data from one dataset (often a polygon layer) is used to clip, or extract, data from a second dataset with a larger extent. For this exercise, you will clip street centerlines using the dissolved buffer created in exercise 2a to identify those streets within 1,000 feet of a school. You will again work in the Buffer & Clip map for this exercise.

Note: If you closed the Buffer & Clip map after completing exercise 2a, reopen it from the Maps folder of the Catalog pane.

Clip streets using the dissolved buffer

1　In the Contents pane, turn on the visibility for Kansas_City_Streets.

A layer of the road network now appears below the dissolved buffer. You will work with both layers in the Clip tool.

2　On the Analysis tab, in the Geoprocessing group, click Tools.

3　In the search text box, type **Clip** and press Enter.

4　In the search results, click Clip (Analysis Tools).

5　In the Clip pane, select Kansas_City_Streets from the Input Features list.

6 Select Buffer_Schools_Dissolve from the Clip Features list.

7 In the Output Feature Class text box, type **Streets_Buffer_Clip**.

8 Click Run.

Streets_Buffer_Clip now appears on the map and in the Contents pane. As can be seen from the attribute table, 2,856 individual street features were clipped from the overall Kansas_City_Streets file. This file can be helpful in helping commanders identify the specific street and address ranges within the immediate surrounding area of public schools.

Note: Results of GIS analyses must often be shared with agency personnel without access to GIS software. In such a case, exporting the attribute table information into a spreadsheet is helpful for dissemination purposes. ArcGIS Pro includes a Table To Excel tool that exports attribute tables directly into Microsoft Excel format. You will work with this tool in chapter 8 ("Automating crime analysis processes").

Exercise 2c: Use the Merge tool

1   In the Catalog pane, in the Maps folder, click Merge & Location to open the map for this exercise.

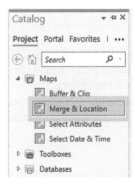

The map displays locations of homicides in Kansas City. Homicide incidents are organized into four layers, one for each quarter of the year: Homicides_KC_Qtr1, Homicides_KC_Qtr2, Homicides_KC_Qtr3, and Homicides_KC_Qtr4.

You will merge the layers into a single homicide layer.

2   In the Geoprocessing pane search text box, type **Merge** and press Enter.

3   In the search results, click Merge (Data Management Tools).

4   In the Merge pane, click the down arrow immediately to the right of Input Datasets.

5   Click each of the four homicide layers.

6   Click Add.

Note: You can also add layers individually from the drop-down lists.

7 For Output Dataset, type **Homicide_Year.**

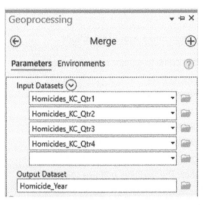

You will not need to adjust any of the settings under Field Map, assuming all the attributes for the layers are identical.

> **Note:** You can still run the tool even if settings in the field map do not match. The downside is that your output layer may contain incomplete attributes. In the worst-case scenario, the data may not merge if there are too many differences in the layers' respective attributes. We recommend ensuring that your layers' attributes match as much as possible when using the Merge tool.

8 Click Run.

The new merged layer now appears in the Contents pane.

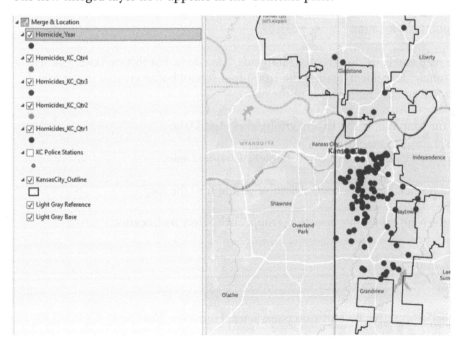

Exercise 2d: Select features by location

Crime analysis highlight 2d: Enhancing reports with location and attribute selections

Many law enforcement agencies report their overall monthly crime to enhance decision-making and improve community and police accountability. Showcasing this data on a map is equally valuable. While agencywide reports are vital, illustrating crime within each geographic boundary (precinct, sector, or district) of an agency decentralizes action plans and allows for local context to influence strategy. In Harris County, we often have to display crime data for special operation units within specific districts. The Select By Location tool allows us to quickly highlight crime incidents that have occurred in areas of interest. We also frequently use Select By Attributes to select groups of incidents that share common characteristics. The flexibility of these selection tools provides analysts with powerful query options.

— *Demetrius DeJean, analyst I, Harris County (Texas) Sheriff's Office*

The Select By Location tool allows you to select features in one layer on the basis of their location relative to features in another layer. This allows you to identify features that share a specific type of spatial relationship. You will use the Select By Location tool to work with the layers you created in the previous exercises.

You will again work in the Merge & Location map for this exercise.

Note: If you closed the Merge & Location map after completing exercise 2c, open it from the Maps folder of the Catalog pane.

Select points by location

The map currently displays the Homicide_Year layer. For this exercise, you will select the homicide incidents within the 1,000-foot school buffer created in exercise 2c. You will add the buffer layer to the map.

1 On the Map tab, in the Layers group, click Add Data.

2 Browse to C:\ModernPolicing\Chapter2\Chapter2.gdb.

3 Double-click Buffer_Schools_Dissolve to add it to the map.

4 On the Map tab, in the Selection group, click Select By Location.

Select By
Location

5 In the Select Layer By Location pane, select Homicide_Year from the Input Features list.

6 Select Intersect from the Relationship list.

7 Select Buffer_Schools_Dissolve from the Selecting Features list.

8 Leave all other options as is.

9 Click OK.

2d

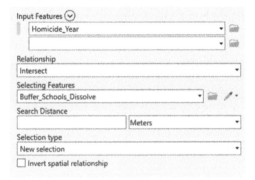

The map now shows that eight of the homicide points fall within the 1,000-foot school buffer, as noted in the lower right.

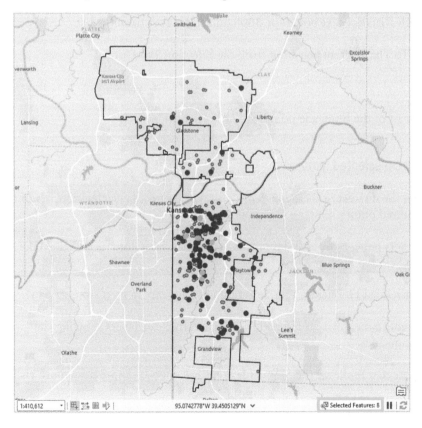

> **Note:** You can also view the number of points selected by clicking the List By Selection button in the Contents pane.

Add features to the current selection

In addition to creating new selections, you can add features to a current selection. For the next step of the exercise, you will add homicides occurring within 3,000 feet of a police station.

1   In the Contents pane, turn off the visibility of Buffer_Schools_Dissolve and turn on the visibility of KC Police Stations.

2   On the Map tab, in the Selection group, click Select By Location.

3   In the Select Layer By Location pane, select Homicide_Year from the Input Features list.

4   Select Intersect from the Relationship list.

5   Select KC Police Stations from the Selecting Features list.

6   In the Search Distance text box, type **3000** and select Feet from the list.

7   Select Add To The Current Selection from the Selection Type list.

8   Click OK.

In the map, 14 of the homicide points are now selected.

Create a layer from selected features

Considering the interest in these incidents, you will create a layer that consists of only the selected points.

1   In the Contents pane, right-click Homicide_Year, click Selection, then click Make Layer From Selected Features.

A new layer named Homicide_Year Selection now appears in the Contents pane.

2   In the Contents pane, turn off Homicide_Year and the four layers showing the quarterly homicides.

The 14 points that make up the Homicide_Year Selection layer are now more visible.

Creating a layer from a selection results in a temporary visual file. In other words, a new GIS file is not created, and thus the selection cannot be imported into another project. You must use the Export Features function to create a stand-alone GIS file from a selection. You will conclude this exercise by converting the selected points into a new file.

3 In the Contents pane, right-click Homicide_Year Selection, click Data, then click Export Features.

4 In the Geoprocessing pane, type **Homicide_Schools_PoliceStations** in the Output Name box.

5 Click OK.

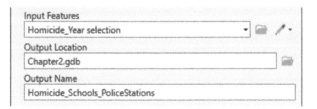

Homicide_Schools_PoliceStations now appears in the Contents pane and the map. Unlike the temporary file (Homicide_Year Selection), this layer is a stand-alone file saved to Chapter2.gdb.

Exercise 2e: Select features by attribute

The Select By Attributes tool allows you to select features on the basis of their under-lying attributes. The tool uses a Structured Query Language (SQL) query to identify (and select) features matching user-generated selection criteria. You will use the Select By Attributes tool to explore the underlying characteristics of unsolved homicide cases in New York City.

Select by attribute value

1 In the Catalog pane, in the Maps folder, click Select Attributes to open the map for this exercise.

This map displays homicides occurring throughout New York City by case status: Closed By Arrest, Open/No Arrest, and Closed Without Arrest. You will use the Select By Attributes tool to isolate the open homicides and explore the features across victim age groups.

2   On the Map tab, in the Selection group, click Select By Attributes.

3   For Input Rows, select Homicide Clearance (NYC).

4   For Selection Type, choose New Selection.

5   Under Expression, click New Expression.

6   In the first expression box, select Disposition.

7   In the operator box, select Is Equal To.

8   In the third box, select Open/No Arrest.

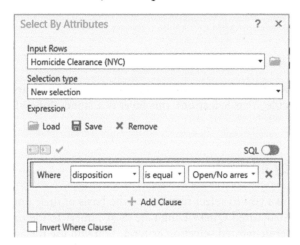

9   Click OK.

Make a layer from selected features

The Open/No Arrest homicides are now selected on the map. You will create a layer consisting of only these incidents.

1   In the Contents pane, right-click Homicide Clearance (NYC), click Selection, then click Make Layer From Selected Feature.

The Homicide Clearance (NYC) selection now appears in the Contents pane. You will change the name of the layer to reflect the nature of the data.

2   Click the Homicide Clearance (NYC) selection and press Enter on your keyboard to open the Layer Properties window.

3 On the General tab, in the Name text box, type **Open Homicides**.

4 Click OK.

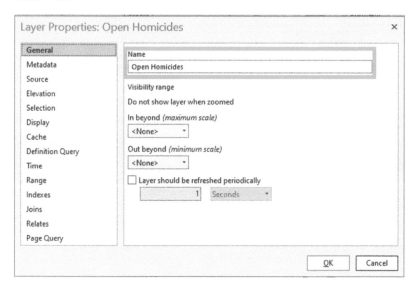

The Open Homicides layer does not include closed cases, so you will remove Closed By Arrest and Closed Without Arrest from the layer's legend in the Contents pane.

5 On the Appearance tab, in the Drawing group, click Symbology.

The Geoprocessing pane displays the primary symbology schema for the Open Homicides layer. The bottom of the window displays symbol styles for all three categories in the original Homicide Clearance (NYC) layer, despite there being only one category (Open/No Arrest) contained within the Open Homicides layer. You will remove the Closed By Arrest and the Closed Without Arrest categories.

6 In the Geoprocessing pane, right-click Closed By Arrest and click Remove.

7 Repeat the previous action for the Closed Without Arrest category.

The Open Homicides layer now contains only those homicides identified as Open/No Arrest.

8   In the Contents pane, turn off the visibility for Homicide Clearance (NYC).

The Open Homicides layer, which includes only the 98 Open/No Arrest incidents, is now more visible on the map.

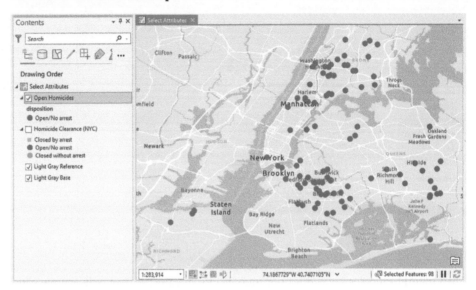

Select features by value range

Now that all the open homicides are isolated, you will use the Select By Attributes tool to explore homicides involving victims of different age groups.

1   On the Map tab, in the Selection group, click Select By Attributes.

2   For Input Rows, click Open Homicides.

3   For Selection Type, choose New Selection.

4   Under Expression, click New Expression.

5   In the first expression box, select Age_Victim.

6   In the operator box, select Is Less Than.

7   In the third box, type **18**.

8   Click Apply.

The nine incidents involving juvenile victims (below the age of 18) are now selected on the map.

Select features through compound expressions

All the attribute selections have used singular query expressions up to this point—a single clause was used to select features. You will use compound expressions involving multiple clauses to select homicides involving victims within different age ranges for the remainder of this exercise.

1 In the Geoprocessing pane, under Expression, click Remove to clear the previous attribute expression.

2 Under Expression, click Add Clause.

3 In the first box, select Age_Victim.

4 In the operator box, select Is Greater Than Or Equal To.

5 In the third box, type **18**.

6 Click Add Clause.

7 In the first box for the new clause, select And.

8 In the second box, select Age_Victim.

9 In the third box, select Is Less Than Or Equal To.

10 In the fourth box, type **25**.

11 Click Apply.

The 29 incidents involving victims between ages 18 and 25 are now selected on the map.

Load saved expressions

You have thus far manually created all clauses in the Select By Attributes tool. You can also load previously created and saved expressions into your project.

1 In the Geoprocessing pane, under Expression, click Remove to clear the previous attribute expression.

2 Under Expression, click Load.

3 Browse to C:\ModernPolicing\Chapter2\Expressions.

4 Select Homicide_age_26-39.exp and click OK.

2e

A new expression that selects homicides with victims between the ages of 26 and 39 is added to the Select By Attributes tool.

5 Click Apply.

The 28 incidents involving victims between ages 26 and 39 are selected on the map.

Create and save an expression

For the final step in this exercise, you will create and save an expression to your computer for future use. You will adjust the current expression to select homicides with victims between ages 40 and 54.

1 In the current expression, delete 26 from the third box and type **40**.

2 Delete 39 from the final box and type **54**.

3 Under Expression, click Save.

4 Browse to C:\ModernPolicing\Chapter2\Expressions.

5 In the Name field, type **Homicide_age_40-54**.

6 Click Save, then click OK.

The 19 incidents involving victims between ages 40 and 54 are now selected on the map.

Exercise 2f: Select features by day and time

The previous exercise introduced you to the process of creating expressions in the Select By Attributes tool for selecting layer features. A common task for crime analysts is selecting data on the basis of its time of occurrence, such as during a period of interest (for example, a 28-day CompStat period) or days of the week (for example, the weekend). Although creating a compound expression in the Select By Attributes tool can perform such a selection, ArcGIS Pro includes a Select By Date And Time tool that enables users to select features occurring within a given time frame.

Select by date and time

1 In the Catalog pane, in the Maps folder, click Select Date & Time to open the map for this exercise.

The Select Date & Time map opens, displaying homicides throughout New York City.

For this exercise, you will select homicides occurring in summer months (June through August) during the afternoon patrol shift (4:00 p.m. to 11:59 p.m.). You will use the Select By Date And Time tool to simultaneously select incidents by these parameters.

2   On the Map tab, in the Selection group, click Select Layer By Date And Time.

Select Layer By
Date and Time

3   In the tool pane, for Input Rows, select Murder & Homicide (NYC).

4   For Selection Type, choose New Selection.

5   For Time Type, choose Single Time Field.

6   For Date Field, select Date.

Five options are present within the tool: Select By Date, Time, Day Of Week, Month, and Year. You will use both the Date and the Time options for the remainder of this exercise.

7   Under Selection Options, check the box next to Date.

The Select By Date line opens at the bottom of the tool pane.

8   Expand the Select By Date heading to view the tool's selection fields.

9   For Date Selection Type, select By Date Range.

10 Adjust the Start Date and the End Date to 6/1/2019 and 8/31/2019, respectively.

You can type information manually or click the clock icon to use a calendar to enter data.

11 Under Selection Options, check the box next to Time.

12 Expand the Select By Time Of Day heading to view the tool's selection fields.

13 Adjust the Start Time and the End Time to 4:00:00 PM and 11:59:00 PM, respectively.

Input Rows

Murder & Homicide (NYC)

Selection Type

New selection

Time Type

Single Time Field

Date Field

DATE

Selection Options
- ☑ Date
- ☑ Time
- ☐ Day of week
- ☐ Month
- ☐ Year

˅ Select by Date

Date Selection Type

By Date Range

Start Date

6/1/2019

End Date

8/31/2019

˅ Select by Time of Day

Start Time

4:00:00 PM

End Time

11:59:00 PM

14 Click Run.

Create a feature layer from selected features

You will conclude this exercise by exporting the 45 selected features in Murder & Homicide (NYC) as a new data layer. In prior exercises, you first created a temporary layer from the selection before exporting features into a stand-alone file. In this step, you will immediately export the selected features from the Murder & Homicide (NYC) layer.

1 In the Contents pane, right-click Murder & Homicide (NYC) and select Attribute Table.

In the attribute table, you see that 45 of the 292 homicides occurred in summer months (June through August) during the afternoon patrol shift (4:00 p.m. to 11:59 p.m.).

2 In the Contents pane, right-click Murder & Homicide (NYC), click Data, then click Export Features.

3 In the Export Features pane, for Output Name, type **Murder_Summer_Afternoon**.

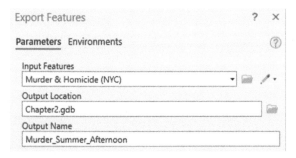

4 Click OK.

Murder_Summer_Afternoon now appears as a stand-alone layer in the Contents pane.

> **Note:** The queries you created in this chapter are merely the beginning of how to use ArcGIS Pro querying tools to select the features you need to answer the toughest questions about your data. For example, assume that information in the New York City Police Department (NYPD) homicide dataset provides the type of weapon used to commit an offense. You could then modify your query in the preceding section to identify weekend homicides that occurred between 8 p.m. and 11 p.m. that also involved a handgun or other firearm. Moreover, a simple yes-or-no designation within the data indicating whether a homicide is gang-related could easily help identify emerging gang rivalries or other similar disputes.

Summary

Chapter 2 introduced you to ArcGIS Pro geoprocessing functions. Geoprocessing is one of the most important aspects of effective GIS, including crime analysis. Your analysis efforts will suffer without the proper access and ability to manipulate data. This chapter's exercises helped you to become proficient in crucial tasks, such as selecting data by location or attributes and merging separate data sources. You are encouraged to read more about model tools and script tools to enhance your data-processing abilities in ArcGIS Pro.

Chapter 3
Creating and editing feature layers

Overview

In this chapter, you will learn how to create, edit, and modify your own GIS data. This involves converting a list of addresses to points on a map as well as drawing and modifying polygon and line features. The techniques outlined in the exercises are the beginning processes of creating rich datasets for crime analysis.

You will acquire the following skills upon completing the exercises in this chapter:
- Geocoding data with x,y coordinates
- Changing the projection of GIS data
- Building address locators for geocoding
- Creating and editing line, point, and polygon features

Download and install the data

Before working on the exercises, you will obtain the necessary data.

1   Go to www.arcgis.com and sign in with your ArcGIS Online account credentials.

2   Type **Modern Policing Using ArcGIS Pro (Esri Press)** in the search box, and click the Groups tab. Make sure that the Only Search In Your Organization option is turned off.

3   Click the Modern Policing Using ArcGIS Pro (Esri Press) group.

4   On the group heading, click Content.

5   Click the chapter 3 file and download it.

6   Locate the zip file you downloaded to your local drive, right-click, and extract it to C:\ModernPolicing.

This will create a folder named Chapter3 in the ModernPolicing folder.

This folder contains an ArcGIS Pro project and data you will use for the exercises in this chapter.

Exercise 3a: Map data with x,y coordinates

Crime analysis highlight 3a: Creating point layers from incident address data

Records Management Systems (RMS) typically contain the raw data that crime analysts use in spatial analysis. Some RMS automatically assign x,y coordinates to each incident, which allows for one-step mapping. When x,y coordinates are not available, crime analysts geocode addresses. Geocoding is the process of assigning geographic coordinates to a location or address. You can either put the entire address together (for example, 123 Liberty Ln, Aqua, NY 12345) or keep each piece of the address in separate fields of the same table. Once you begin the geocoding process, you will determine the fields within your data that hold the addresses you want to map. Next, you determine which coordinate system to apply to the map projection. A widely used coordinate system is the World Geodetic System (WGS) 1984, which is the standard reference for the Global Positioning System (GPS). Commonly, local data used by crime analysts will be projected in a coordinate system that reflects their agency's jurisdiction, such as New York State Plane. You can also change the coordinate system of your data after geocoding by transforming the layer's projection.

— *Sarah Eason, crime research analyst, North Country (New York) Crime Analysis Center*

Display x,y data

1   Open Chapter3.aprx from C:\ModernPolicing\Chapter3.

The project shows a map named x,y. The map displays the city boundaries for Jersey City, New Jersey. In this exercise, you will prepare the feature layers necessary to track the change in assaults following the creation of a business improvement district. The first feature layer will be generated from a table containing x,y coordinates of assault events.

2   In the Contents pane, right-click JC_Assault_Pre and click Open.

This table lists the assault incidents that occurred during the year prior to the establishment of the business improvement district. The table includes the x,y coordinates for each incident.

X_Coor	Y_Coor
607867.265739	687300
608925.60647	684859
606235.587244	687993
615382.547297	691114
615382.547297	691114
621486.33463	688020

3   In the Contents pane, right-click JC_Assault_Pre and click Display XY Data.

4 In the Display XY Data pane, type **Pre_Assault** in the Output Feature Class text box.

5 In the X Field list, select X_Coor.

6 In the Y Field list, select Y_Coor.

7 In the Coordinate System list, select Jersey City Border to ensure that the new layer shares the same projection as the basemap layers you will work with in future exercises (NAD 1983 State Plane New Jersey FIPS 2900 Feet).

8 Click OK and ignore any warnings.

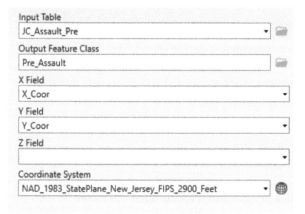

The newly created Pre_Assault layer now appears in the Contents pane.

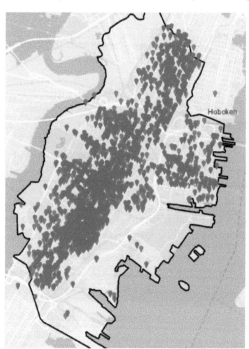

Change feature layer projection

When creating the preintervention assault feature layer, you manually set the projection to match the basemaps that will be used in future exercises. The remaining exercises in this chapter will also incorporate assault data from the one-year postintervention period. You will ensure that the projection of the postintervention data matches the basemaps.

1 In the Catalog pane, open the Databases folder and open Chapter3.gdb.

2 Right-click JC_Assault_Post and click Add To Current Map.

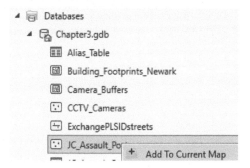

3 In the Contents pane, double-click JC_Assault_Post to open the Layer Properties window.

4 In the Layer Properties window, click Source, scroll down to the Spatial Reference section, and expand it.

JC_Assault_Post uses the WGS84 geographic coordinate system, which differs from the New Jersey State Plane projection used for Pre_Assault.

5 Close the Layer Properties window.

6 In the Contents pane, right-click JC_Assault_Post, and click Attribute Table.

Despite being in the same city as Pre_Assault, JC_Assault_Post has very different x,y coordinate values. The x-coordinates all have negative values (as opposed to the positive values for Pre_Assault) and both the x- and y-coordinates have only two numeric values to the left of the decimal point (instead of six for the Pre_Assault coordinates). This is a by-product of the WGS84 coordinate system, which measures coordinates as longitude and latitude locations in decimal degrees rather than feet.

X_Coor	Y_Coor
-74.069885	40.721825
-74.077278	40.701897
-74.077278	40.701897
-74.077278	40.701897
-74.039047	40.751003

You will now create a new version of postintervention assaults that is projected in the New Jersey State Plane coordinate system.

7 On the Analysis tab, in the Geoprocessing group, click Tools.

8 In the Geoprocessing pane tool search box, type **Project**.

9 Double-click Project (Data Management Tools) to open it.

10 In the Project tool pane, for Input Dataset Or Feature Class, select JC_Assault_Post.

11 For Output Dataset Or Feature Class, type **Post_Assault**.

12 For Output Coordinate System, select Pre_Assault.

This will ensure that Post_Assault is projected in the same coordinate system as Pre_Assault.

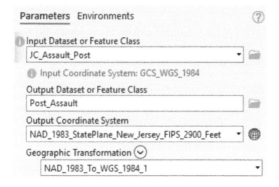

13 Click Run.

Update x,y coordinates in an attribute table

The Post_Assault Layer Properties window lists New Jersey State Plane as the coordinate system. However, changing the projection of a feature layer does not change the attribute fields. The x,y coordinates listed in the attribute table still reflect the WGS coordinate system.

You will conclude this exercise by recalculating the x,y coordinates so that they match the New Jersey State Plane coordinate system.

1 In the Post_Assault attribute table, right-click the X_Coor field, and select Calculate Geometry.

2   Under Geometry Property, set the following parameters:
- For the upper-left Target Field, select X_Coor.
- For the upper-right Property, select Point x-coordinate.
- For the lower-left Target Field, select Y_Coor.
- For the lower-right Property, select Point y-coordinate.

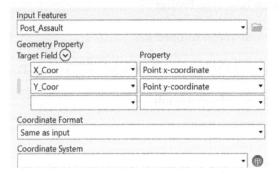

3   Click OK.

The Post_Assault attribute table displays x,y coordinates in the New Jersey State Plane format.

X_Coor	Y_Coor
611337.157421	688148.090791
609323.018733	680878.447522
609323.018733	680878.447522
609323.018733	680878.447522
619828.543654	698820.93854

Exercise 3b: Create address locators and geocode address data

The prior exercise used x,y coordinates to geocode the address data. If you have access to only a list of addresses, you must first create an address locator. Address locators require the use of a reference file, such as a street centerline or parcel layer.

Create an address locator

1   In the Catalog pane, in the Maps folder, click Geocode to open the map for this exercise.

You must understand the structure of your reference file (in this case, the Newark street centerline) before creating an address locator for geocoding. From viewing the attribute table, you can see that the Newark Streets layer is formatted as dual street ranges: numeric address ranges are provided for both the left and the right side of each street segment. The street name, type, prefix, and suffix are contained in separate fields. You will use this information to create the appropriate address locator for your data.

You will geocode the locations of the Newark Police Department video surveillance cameras using the address locator you create in this exercise. It is helpful to view the address table at the outset to understand the nature of the data. The surveillance camera locations are contained in the Cameras_Newark.csv table at the bottom of the Contents pane. From viewing the address table, you see that all camera addresses are contained within a single field (named Address).

Address	Precinct
12th Ave & S 8th St	1st
15th Ave & S. 15th St	4th
16th Ave & S. 15th St	4th
16th Ave & S. 18th St	4th
18th Ave & Livingston St	1st

It is also evident from further review that certain camera locations are denoted by alias place-names, such as apartment complexes (for example, 27 Baxter Terr Homes, 25 Colonnade Homes), rather than a street address (for example, 27 Nesbitt St., 25 Clifton Ave.). You will incorporate an alias table into your address locator to assign each place-name to a street address contained within the reference file.

OBJECTID *	ID	NAME	Street
1	3985	UMDNJ	Bergen St.
2	2820	Colonnade Homes	Clifton Ave.
3	2407	Sacred Heart	Ridge St.
4	3174	Baxter Terrace Homes	Nesbitt St
5	4164	Prudential Tower	Broad St.

2   On the Analysis tab, click the Tools button.

3   In the Geoprocessing pane, click the Toolboxes tab.

4   From the list of Geoprocessing toolboxes, click Geocoding Tools to expand it.

5   Click the Create Locator tool.

6   In the tool pane, for County Or Region, select United States.

7   From the Primary Table(s) list, select Newark_Streets.

8   For Role, select Street Address.

Newark Streets will be the reference file that guides the geocoding process. Leave this as is. In the Field Map area, ArcGIS Pro needs all the necessary address components in the reference file to build the address locator. ArcGIS Pro can identify the role of each field based on its name (for example, L_F_ADD is the Left House Number From address range).

9 Confirm the following field-mapping values:
- Street Join ID = ID

> **Note:** The address locator would still function without a Street Join ID being specified. However, mapping this field is necessary when the address locator includes an alias table.

- Left House Number From = L_F_ADD
- Left House Number To = L_T_ADD
- Right House Number From = R_F_ADD
- Right House Number To = R_T_ADD
- Prefix Type = PREFIX
- Street Name = NAME

> **Note:** A number of records in the Newark Streets layer do not have any values in the NAME field. Many real-world datasets will have such missing or incomplete information. The geoprocessing tool warning allows you to consider this.

- Suffix Type = TYPE

10 From the Language Code list, select English.

Add an alias table to the address locator

You will now load the alias table from the hard drive into the address locator.

1 In the Geoprocessing pane, click the Optional Parameters tab.

2 From the Alternate Name Tables list, select Alias_Table.

3 For Role, select Alternate Street Name.

4 Confirm the following field mapping values:
- Street Join ID = ID
- Street Name = NAME

5 In the Output Locator box, type **Newark_Streets_Locator**.

6 Click Run.

Parameters Environments   ⑦

Country or Region

| United States | ▼ |

Primary Table(s) ⊙

| Newark Streets | ▼ | 📁 | Street Address | ▼ |
| | ▼ | 📁 | | ▼ |

Field Mapping

Role: Street Address

Field Name	Alias Name
Street Join ID	ID ▼
*Left House Number From	L_F_ADD ▼
*Left House Number To	L_T_ADD ▼
*Right House Number From	R_F_ADD ▼
*Right House Number To	R_T_ADD ▼
Left Parity	<None> ▼
Right Parity	<None> ▼
Prefix Direction	<None> ▼

Output Locator

| Newark_Streets_Locator | 📁 |

Language Code

| English | ▼ |

∨ **Optional parameters**

Alternate Name Tables ⊙

| Alias_Table | ▼ | 📁 | Alternate Street Name | ▼ |
| | ▼ | 📁 | | ▼ |

Alternate Data Field Mapping

Role: Alternate Street Name

Field Name	Alias Name
*Street Join ID	ID ▼
Prefix Direction	<None> ▼
Prefix Type	<None> ▼
*Street Name	NAME ▼
Suffix Type	<None> ▼
Suffix Direction	<None> ▼
Full Street Name	<None> ▼
Language Code	<None> ▼

Custom Output Fields

| |

Precision Type

| Global High | ▼ |

3b

The newly created address locator is added to the Locators folder in the Catalog pane.

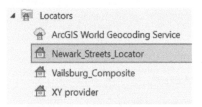

> **Note:** Street Address is one of many types that can be used in an address locator. ArcGIS Pro allows for a wide range of reference dataset geometry to be incorporated in the geocoding process. For more information on primary address locator roles and associated reference data, see "Primary locator roles" in ArcGIS Pro Help.

Customize address locator properties

The Locator Properties window allows you to make changes to various aspects of the address locator. You will adjust some of the default properties before using the address locator to improve geocoding quality. First, you will disable the Match Out Of Range property, which places points on the end of a street segment when the address is slightly outside the house number range. Second, you will increase the minimum match score to 85. The match score is a value between 0 and 100, with a perfect match yielding a score of 100. A score of 85 provides a balance between achieving perfect matches and setting a score so low that some addresses are potentially matched incorrectly. Changing the Match Out Of Range property and the minimum match score allows for a more cautious approach that gives users more control over how cases are matched. Last, address locators in ArcGIS Pro will not match addresses without accompanying zone information, such as a zip code or city. This option is necessary when mapping data containing addresses from multiple jurisdictions. However, this option may be impractical when working with data from a single jurisdiction because police data systems may not contain zone information. This setting needs to be adjusted in such cases.

1 In the Catalog pane, under Locators, right-click Newark_Streets_Locator, and click Locator Properties.

2 In the Locator Properties window, click Geocoding Options.

3 For Match Out Of Range, select No.

4 For Minimum Match Score, delete 60 and type **85**.

About the locator	Match out of range
Reference Data Tables	No ▾
Alternate Name Tables	
Input fields	Minimum match score
Output fields	85
Geocoding options	Minimum candidate score
Performance	60

5 For Match With No Zones, confirm it is set to Yes, and click OK.

Geocode an address table

1 In the Contents pane, right-click Cameras_Newark.csv, and click Geocode Table.

The Geocode Table tool opens.

2 In the tool pane, click Start.

The first step asks whether the addresses in your table are contained in one field or more than one field. In this case, the addresses are kept in one field (address).

3 From the list, select One Field, and click Next.

4 For Input Locator, select Newark_Streets_Locator, and click Next.

5 For Data Field, select Address, and click Next.

6 For Output, name the file **Newark_Cameras**, and click Next.

7 Under Category, check the box for Addresses, and click Finish.

8 Confirm your settings and click Run.

Review and rematch addresses

The geocoding results window now shows that one of the 141 surveillance camera locations was unmatched (and three were tied). You will rematch this address.

1 In the geocoding results window, click Yes to start the rematch process.

The unmatched address, ITB & W. Kinney St., now appears in the Rematch Address pane. ITB is short for Irvine Turner Blvd., which is why the address was not matched to the Newark street centerline.

2   For Single Line Input, type **Irvine Turner Blvd. & W. Kinney St.,** and press Enter.

A candidate address appears at the bottom of the Geoprocessing pane. The map view also displays the location of this address. The candidate address has a match score of 100, meaning that it is identical to the address currently in the text box.

3   In the Rematch Address pane, click the green check mark to match the candidate address.

The address is now matched to the candidate address.

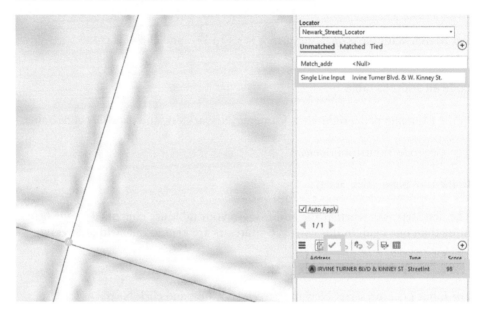

4   Click the Save Edits button in the Geoprocessing pane, confirm the saved edits by noting that no further candidates have been found, and close the tool.

> **Note:** When rematching addresses, you can also select x,y coordinates from the map by using the Pick From Map tool. This tool allows you to match addresses by clicking a location within the map view. You will use this tool in exercise 3c. Several improvements to geocoding accuracy were introduced in ArcGIS Pro. Four addresses in the Cameras_Newark.csv table do not have spaces between the street name and street type (for example, EdisonPl). In the legacy ArcGIS software, this error would have resulted in a low match score and required manual rematching for geocoding to occur successfully. ArcGIS Pro automatically matches these addresses to the reference file without requiring manual rematching. Each of the aforementioned addresses achieved match scores of over 99 percent in this exercise.

You will conclude this exercise by inspecting the attribute table of the newly geocoded layer to gain a sense of what fields are added to a table by the address locator.

5 Open the Newark_Cameras attribute table. (**Hint:** In the Contents pane, right-click Newark_Cameras, and click Attribute Table.)

As part of the geocoding process, numerous fields are added to the attribute table, including match status, match score, match address, address type (street address or intersection), as well as individual components of the address (number range, and so on). You can change the geocoding options of the address locator so that it does not include certain fields, if you prefer (see step 3 in the "Customize address locator properties" section). You can also determine which records were matched with the assistance of the alias table from within the attribute table. For example, the address locator matched the address Prudential Tower to 655 Broad St. Given that the reference file does not contain any street segments named "Prudential Tower," this point would not have geocoded without the alias table.

3b

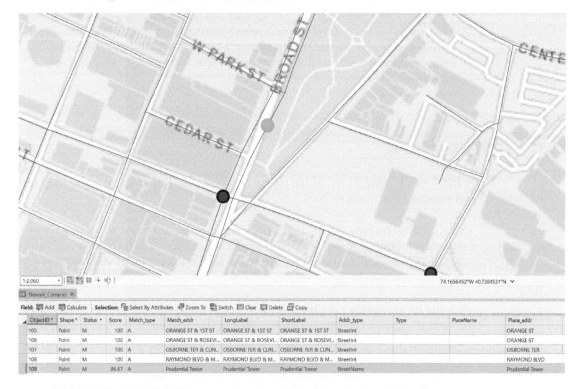

ObjectID *	Shape *	Status *	Score	Match_type	Match_addr	LongLabel	ShortLabel	Addr_type	Type	PlaceName	Place_addr
105	Point	M	100	A	ORANGE ST & 1ST ST	ORANGE ST & 1ST ST	ORANGE ST & 1ST ST	StreetInt			ORANGE ST
106	Point	M	100	A	ORANGE ST & ROSEVI...	ORANGE ST & ROSEVI...	ORANGE ST & ROSEVI...	StreetInt			ORANGE ST
107	Point	M	100	A	OSBORNE TER & CLIN...	OSBORNE TER & CLIN...	OSBORNE TER & CLIN...	StreetInt			OSBORNE TER
108	Point	M	100	A	RAYMOND BLVD & M...	RAYMOND BLVD & M...	RAYMOND BLVD & M...	StreetInt			RAYMOND BLVD
109	Point	M	86.67	A	Prudential Tower	Prudential Tower	Prudential Tower	StreetName			Prudential Tower

Note: In addition to assigning addresses to place-name aliases, alias tables can be used to rectify common mistakes in street addresses, which can minimize the amount of time needed to manually review or match unmatched addresses.

Exercise 3c: Edit line and point features

Research highlight 3c: Editing street segments and points for spatial analysis

Crime-and-place research has increasingly focused on microlevel units of analysis to measure crime, disorder, and the effect of police interventions. Microunits, such as street segments, street intersections, and street addresses, offer distinct benefits over larger geographic units (Weisburd et al. 2009). Using large geographic areas to study crime may lead to the false conclusion that observed patterns equally apply to the smaller units of which they consist—a problem referred to as the *ecological fallacy* (Johnson et al. 2009). Empirical research has found that microlevel units of analysis explain a larger share of the spatial variability of crime than either meso- or macrolevel units (Schnell et al. 2017; Steenbeek and Weisburd 2016). Given the importance of microlevel units, crime analysts may need to periodically edit basemaps to ensure that street segments accurately capture underlying crime patterns throughout a city. The precise placement of crime points is also important in light of recently developed statistical tests meant to detect changes in spatial point patterns (Andresen and Malleson 2011). The ArcGIS Pro Colocation Analysis tool, covered in chapter 7 ("Using spatial statistics to identify spatial relationships"), uses such a point-pattern test.

Merge line features

1   In the Catalog pane, in the Maps folder, click Lines & Points to open the map for this exercise.

This map displays two group layers, one named Jersey City and one named Newark. Group layers control drawing options for all the sublayers in the group and help to organize the content in maps. You will work with the Jersey City data for this portion of the exercise.

Visibility for the Jersey City layer is turned on. However, aside from the city border, only the streets layer is currently visible in the map view. The street intersection points are not visible owing to the visibility range currently assigned to the layer. Users can apply a visible range to layers so that features appear in the map view only at certain extents. In the current exercise, the visible range prevents the street intersections from appearing at the citywide extent. You will view the visibility parameters currently assigned to the intersections.

2   In the Contents pane, click Street Intersections.

A series of feature layer contextual tabs appears on the ribbon.

3   Click the Appearance tab.

The Visibility Range group on the Appearance tab lists the current visibility range for street intersections, which is restricted beyond the scale of 1:5,000 (Out Beyond value). You will zoom in beyond this limit to view the intersections.

4 Go to the Line Edit 1 bookmark. (**Hint:** On the Map tab, click Bookmarks, and select Line Edit 1.)

The map view now shows Cator Ave. between Spring St. and Fowler Ave. The intersection points were generated at every point in Jersey City where multiple street segments intersect (that is, street corners). A point was generated on Cator Ave. even though it is a single street segment. This point was created because the single segment of Cator Ave. consists of two separate lines. You will merge these two lines into a single segment.

5 On the Map tab, in the Selection group, click Select, and choose the Rectangle tool.

6 Select the Cator Ave. line to the northwest of the red intersection point.

7 While pressing Shift on your keyboard, select the Cator Ave. line to the southeast of the red intersection point.

With both lines selected, you will merge them into a single feature. You will note the address ranges of the separate features to assign the correct address range to the newly merged feature in the next section.

8 Open the attribute table for JerseyCity_Streets.

9   Click Show Selected Records at the bottom left.

In the attribute table, you can see that the selected features have their own unique numeric address information. It is important to note this information so that you can assign the proper address range to the merged street segment.

> **Note:** Single street segments are often represented as multiple lines for the purpose of geocoding accuracy. For example, building addresses may be evenly placed at the beginning of a street segment and more closely distributed toward the end. Drawing single street segments with multiple lines allows for numeric address ranges to better reflect this reality. Therefore, merging street segments may compromise geocoding accuracy. As such, it is good practice to maintain two versions of street segment files, one structured to maximize geocoding and another structured for the purpose of spatially aggregating points for spatial analysis.

10   On the Edit tab, in the Tools group, click Merge.

3c

The Geoprocessing pane now displays the Merge tool. The two selected line features appear in the Features To Merge section. The selected feature indicated with Preserve in parentheses will have its attributes applied to the new feature.

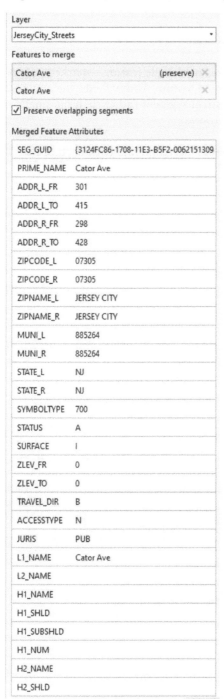

Layer

JerseyCity_Streets ▾

Features to merge

| Cator Ave | (preserve) ✕ |
| Cator Ave | ✕ |

☑ Preserve overlapping segments

Merged Feature Attributes

SEG_GUID	{3124FC86-1708-11E3-B5F2-0062151309
PRIME_NAME	Cator Ave
ADDR_L_FR	301
ADDR_L_TO	415
ADDR_R_FR	298
ADDR_R_TO	428
ZIPCODE_L	07305
ZIPCODE_R	07305
ZIPNAME_L	JERSEY CITY
ZIPNAME_R	JERSEY CITY
MUNI_L	885264
MUNI_R	885264
STATE_L	NJ
STATE_R	NJ
SYMBOLTYPE	700
STATUS	A
SURFACE	I
ZLEV_FR	0
ZLEV_TO	0
TRAVEL_DIR	B
ACCESSTYPE	N
JURIS	PUB
L1_NAME	Cator Ave
L2_NAME	
H1_NAME	
H1_SHLD	
H1_SUBSHLD	
H1_NUM	
H2_NAME	
H2_SHLD	

11 At the bottom of the tool pane, click Merge.

3c

12 On the Edit tab, click Save.

The two separate line features are now merged into a single feature.

Update the attribute table for the merged line feature

The Merge Features tool assigns the attributes of a single feature—marked as (preserve) during the merge process—to the merged feature. You will edit the attribute table to reflect the address range of the full street segment rather than the single street segment preserved during the merge process.

1 For the merged feature (selected in the attribute table), double-click each of the address range features, and type the following values:
- ADDR_L_FR: **299**
- ADDR_L_TO: **415**
- ADDR_R_FR: **296**
- ADDR_R_TO: **428**

2 On the Edit tab, click Save, and clear the selected features.

Split a line feature

In the previous portion of this exercise, you joined two line features to represent a single street segment. You will now split a single line feature to represent two separate street segments.

1 Open the Line Edit 2 bookmark.

The map view shows Dudley St. between Van Vorst St. and Washington St. Dudley St. is represented by a single line feature. Dudley St. should consist of two street segments, one extending from Van Vorst St. to Warren St. and one extending from Warren St. to Washington St. You will split the line feature to create these separate street segments.

2 On the Map tab, in the Selection group, click the Select By Rectangle tool.

3 In the map view, select Dudley St.

4 On the Edit tab, in the Tools group, click Split.

5 Click the Warren St. street segment to the north of Dudley St.

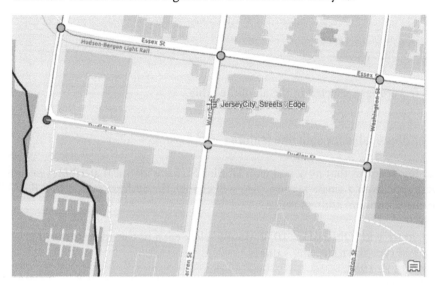

6 Move your pointer southward, and click the Warren St. segment south of Dudley St.

7 Double-click to complete the split process.

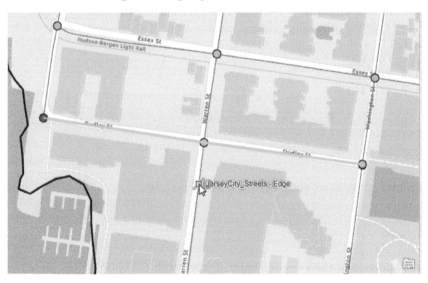

Dudley St. is now split into two separate line features.

8 Open the attribute table for JerseyCity_Streets, and click Show Selected Records.

OBJECTID *	Shape *	SEG_GUID	PRIME_NAME ▲	ADDR_L_FR	ADDR_L_TO	ADDR_R_FR	ADDR_R_TO
1070	Polyline M	{2F4BC868-1708-11E3-...	Dudley St	198	100	199	101
3963	Polyline M	{2F4BC868-1708-11E3-...	Dudley St	198	100	199	101

The two features are selected in the attribute table.

Update the attribute table for split line features

The two line features that now make up Dudley St. have identical attributes because they were split from a single feature. You will change the address ranges to reflect each segment's unique address range.

1 For the feature with ObjectID = 1070, double-click each of the address range features, and type the following values:
- ADDR_L_FR: **198**
- ADDR_L_TO: **126**
- ADDR_R_FR: **199**
- ADDR_R_TO: **127**

2 For the feature with ObjectID = 3963, double-click each of the address range features, and type the following values:
- ADDR_L_FR: **124**
- ADDR_L_TO: **100**
- ADDR_R_FR: **125**
- ADDR_R_TO: **101**

OBJECTID *	Shape *	SEG_GUID	PRIME_NAME ▲	ADDR_L_FR	ADDR_L_TO	ADDR_R_FR	ADDR_R_TO
1070	Polyline M	{2F4BC868-1708-11E3-...	Dudley St	198	126	199	127
3963	Polyline M	{2F4BC868-1708-11E3-...	Dudley St	124	100	125	101

3 On the Edit tab, click Save, and clear the selected features.

Edit point features

For the final portion of the exercise, you will work with data in Newark, New Jersey, to edit existing point features.

1 Open the Point Edit bookmark.

2 Turn off the visibility for the Jersey City group layer and turn on the visibility for the Newark group layer.

The map currently displays points and building footprints in the immediate vicinity of Bradley Court, a housing complex in the Vailsburg section of Newark, New Jersey.

The points represent each address on North Munn Ave. between Mountainview Ave. and Tremont Ave. The points do not accurately reflect the spatial distribution of addresses as currently placed. The left side of North Munn Ave. consists of buildings immediately adjacent to the street centerline, which are accurately represented by the current points. The right side of North Munn Ave., on the other hand, has an address range that corresponds to 10 buildings in the Bradley Court complex, many of which are a substantial distance from the street centerline. As such, the points corresponding to the right side of North Munn Ave. poorly reflect the spatial distribution of the addresses. You will conclude this exercise by repositioning these points accordingly.

Manually move points

The most straightforward method of repositioning points is to manually move them in the map view.

1   In the Contents pane, click N_Munn_Addresses_Geocoded.

2   On the Edit tab, in the Tools group, click Move.

3c

3 Click the northernmost point to the right side of North Munn Ave.

This point corresponds to 90 North Munn Ave. You will now reposition the point to the building footprint corresponding to 88–90 North Munn Ave.

4 With the point for 90 N. Munn Ave. selected, hold down the left mouse button and drag the point to the center of the 88–90 North Munn Ave. building footprint.

5 Double-click to finalize the new spatial location of the point.

6 On the Edit tab, click Save.

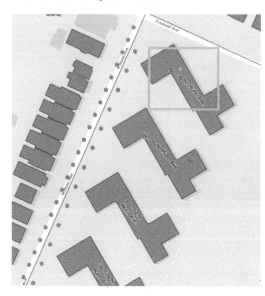

3c

Rematch point addresses

Another way to reposition points is to use the Rematch Addresses tool, which opens following geocoding. You can access the tool for previously geocoded points from within the Contents pane.

1 In the Contents pane, right-click N_Munn_Addresses_Geocoded, click Data, and click Rematch Addresses.

The Rematch Addresses tool opens in the Geoprocessing pane. The tool lists the address for each individual point contained in the feature layer. The first of the 54 addresses (21 N. Munn Ave.) appears in the tool pane. You are interested in rematching only the evenly numbered addresses in this instance, which begin with feature 36.

2 Click the right arrow in the middle of the tool pane to advance to case 36 / 54 (46 N. Munn Ave.).

3 Click the Pick From Map button.

4 Move your pointer over the building footprint for 46–50 N. Munn Ave and click.

The x,y coordinates of the point picked from the map now appear as the first candidate address.

5   Click the green check mark to assign the x,y coordinates to the point.

6   Repeat steps 2 through 5 for three more of the evenly numbered addresses, choosing the correct building outline for each address.

7   Click the Save Edits button after all points have been rematched.

> **Note:** Another option for repositioning the N. Munn Ave. address points is to use a different address locator to geocode the original address table, given that the street centerline address locator inappropriately placed the evenly numbered addresses. One option is to create a separate address locator using the building footprints as the reference data. A composite address locator that combines the building footprint and street centerline locators could then be created to ensure that N. Munn Ave's evenly numbered addresses are plotted to building footprints while the oddly numbered addresses were plotted to street centerlines. As an example, such a composite address locator (Vailsburg_Composite) is available in C:\ModernPolicing\Chapter3. You can use this address locator to geocode N_Munn_Addresses.csv in C:\ModernPolicing\Chapter3\Tables.

Exercise 3d: Create polygon features

Research highlight 3d: Business improvement districts as drivers of community policing

Policing scholars widely consider geographic decentralization of police commands as a fundamental element of community policing (Skogan 2019). Decentralized strategies commonly involve police soliciting the insight and active participation of community members to customize prevention efforts to the underlying nature of crime in the target area. Business improvement districts (BIDs) have emerged as a popular model to foster community policing in commercial areas by shifting crime prevention from solely a police responsibility to an active collaboration between public and private stakeholders. This is particularly the case when police establish crime prevention interventions in conjunction with BIDs (Piza et al. 2020). BIDs commonly incorporate a self-imposed financing mechanism on all business and property owners within a given district, with the generated revenue funding key neighborhood services (Brooks 2008). BIDs have been associated with decreased levels of crime compared with similar areas absent BIDs (Brooks 2008; Hoyt 2004; MacDonald et al. 2009). BIDs have been shown to be cost effective, with every $10,000 spent by a BID on crime control strategies in Los Angeles generating over $200,000 in societal benefits achieved through reduced crime levels (Cook and MacDonald 2009). BIDs have the potential to become a key component of modern community-policing efforts.

3d

Create a feature class

1   In the Catalog pane, in the Maps folder, double-click Polygon, Create.

This map shows all street segments falling within the Exchange Place Special Improvement District, a Jersey City, New Jersey, BID established in 2017.

For this exercise, you will create a polygon feature to represent the boundaries of the Exchange Place Special Improvement District.

2   In the Catalog pane, open the Databases folder.

3   Right-click Chapter3.gdb, click New, and click Feature Class.

The Create Feature Class tool opens in the pane.

4   In the tool pane, for Name, type **SID_ExchangePl.**

5 Because this is not a route or a 3D feature, make sure that the boxes in the Geometric Properties section (m-values and z-values) are not checked.

6 Click Next.

The Fields pane displays the fields of the feature, which will be visible in the attribute table.

7 Click Add A New Field, and type **Zone** for the new field name.

8 For Data Type, leave Text, and click Next.

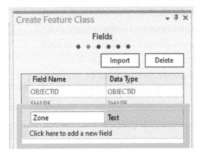

The Spatial Reference panel assigns the projection of the new feature class. In the XY Coordinate Systems Available section, under Layers, the coordinate systems are listed. These coordinate systems are used by the layers in the map. You will ensure that the new feature class uses the same projection as the Exchange Pl SID Streets layer.

9 In the XY Coordinate Systems Available section, under Layers, click NAD 1983 StatePlane New Jersey FIPS 2900 (US Feet), and click Next.

The remaining pages in the Create Feature Class tool pane relate to the tolerance, resolution, and storage configuration of the new polygon feature. You will use the default settings for each.

10 Click Finish to create the feature class.

Draw a polygon feature

The SID_ExchangePl feature class is now present in the project's geodatabase and should be visible in the map view.

You will set the snapping preferences before drawing the polygon. The Snapping tool manages how the pointer locks onto other features as you hover near them. This helps to accurately align the newly created features with other features in the map.

1   On the Edit tab, click the Snapping tool arrow.

2   In the Snapping options, turn on snapping and adjust the remaining snapping options so that Point, Endpoint, and Edge Snapping are the only options turned on.

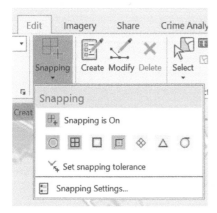

Your polygon will now snap to the edges of underlying map features. For this exercise, edge snapping will ensure that the SID_ExchangePl polygon feature aligns with the underlying street segments that are part of the Special Improvement District.

3   In the Contents pane, select SID_ExchangePl.

4   On the Edit tab, click Create.

> **Note:** In ArcMap, users were required to start an editing session to use the various editing tools. Editing is immediately available in ArcGIS Pro by clicking any of the tools on the Edit tab.

5 In the Create Features tool pane, double-click SID_ExchangePl.

6 Click the blue right arrow to activate this layer's template for editing.

You will use the Trace tool to ensure that the polygon corresponds to the boundaries of the SID streets.

7 In the Active Template pane, click the Trace tool.

You will also update the attributes for the new feature.

8 For the Zone attribute, type **Edison Place**.

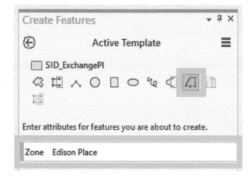

9 With the Trace tool activated, move your pointer to the upper-left corner of the Exchange_Pl_SID_Streets layer.

10 Hold down the left mouse button and trace the edge of the outermost streets in the SID.

11 When you have finished tracing the polygon, double-click to finish the digitizing process.

12 Click Save.

The newly created polygon is now a feature in the SID_ExchangePl layer and can be confirmed by viewing the attribute table.

> **Note:** It is helpful to deactivate the editing tools and activate the map navigation functions when you finish digitizing. If you do not do so, you run the risk of making further, inadvertent edits to the map features when attempting to zoom or pan the map.

Exercise 3e: Edit polygon features (part 1)

Research highlight 3e: Creating viewsheds to evaluate video surveillance camera systems

The use of closed-circuit television (CCTV) surveillance cameras has grown exponentially in recent times. A key challenge in analyzing the effect of CCTV on crime is exactly how to measure the geography of camera target areas. While researchers have most often used large geographies such as neighborhoods or police districts in CCTV research, these areas are poor representations of camera coverage. This is because CCTV cameras are only able to see limited distances, meaning that most of a neighborhood or police district falls outside of the cameras' gaze. Another common approach is the use of circular buffer areas drawn around CCTV cameras to approximate the line of sight. Buffer areas assume a 360-degree, unobstructed line of sight, which rarely occurs in a real-world environment (Piza et al. 2014a). Given these limitations, recent research has used GIS to digitize CCTV viewsheds that estimate each camera's actual line of sight (Caplan et al. 2011; Lim and Wilcox 2017; McLean et al. 2013; Piza et al. 2014a; Ratcliffe et al. 2009; Ratcliffe and Groff 2018). Certain studies have digitized the live feeds of CCTV cameras to digitize the areas visible to cameras (Piza et al. 2014a; Ratcliffe et al. 2009) while others have used aerial imagery and GIS basemap features to estimate CCTV camera viewsheds (Caplan et al. 2011; Piza 2018). In both approaches, GIS editing functions allow researchers to create units of analysis necessary to measure CCTV effect on a microlevel.

3e

Add features to a feature class

Whereas the prior exercise involved creating a feature class from scratch, this exercise requires you to create features within an existing feature class.

1   In the Catalog pane, in the Maps folder, click Polygon, Edit to open the map.

The map currently shows the seven CCTV surveillance cameras installed in the Fourth Police Precinct of Newark, New Jersey. Five CCTV cameras fall within a polygon feature class denoting the estimated viewshed of each camera. The viewsheds account for visible obstructions presented by nearby buildings. The buffers represent the median maximum visible extent of Newark's CCTV cameras (423 feet) as measured in prior research (Piza et al. 2014a; Piza 2018).

Note: The viewsheds are symbolized with a transparency of 50 percent, allowing you to see the underlying basemap through the polygons. You can adjust the transparency on the Appearance tab.

Two of the CCTV cameras do not currently have any accompanying viewshed. You will create viewsheds for these cameras in this exercise.

2   Open the Create Viewshed 1 bookmark.

The camera on 18th Ave. & S. 18th St. appears in the map view. You will now create a viewshed feature for this camera.

3 On the Edit tab, in the Features group, click Create.

4 In the Create Features pane, click Viewsheds.

5 Click the blue right arrow to activate this layer's template for editing.

6 Confirm that the Polygon tool is active.

7 For the Camera attribute, type **18th Ave. & S. 18th St.**

8 In the map view, move your pointer over the bottom corner of the nearest building to the northeast of the camera, and click to add a vertex.

You will now move your pointer within the buffer to draw the camera viewshed, excluding any areas within the buffer that are obstructed from the camera's line of sight. Click often to add vertices as you move the pointer to draw the viewshed.

9 When you have finished creating the polygon, double-click to finish the digitizing process.

To further practice drawing polygon features, open the Create Viewshed 2 bookmark and repeat steps 1 through 9, typing **395 Avon Ave.** for the Camera attribute.

Edit vertices of a polygon feature

You will now edit a viewshed that was drawn incorrectly. Before moving forward, you want to ensure that other layers are not affected while you work on the viewsheds. Because the buffers and viewsheds overlap, you run the risk of inadvertently editing the buffer while working on the viewshed.

1 In the Contents pane, click List By Editing, and uncheck the box for Camera_Buffers.

The Camera_Buffers layer can no longer be edited.

2 Open the Edit Viewshed bookmark.

The map displays a viewshed extending beyond the buffer (that is, the visibility range) to the east. Additionally, the viewshed extends past a building to the northwest, an area visibly obstructed from the camera.

3 On the Edit tab, in the Tools group, click Edit Vertices.

4 In the map view, click the Viewsheds layer.

The vertices of the feature, which were created when the viewsheds were originally drawn, now appear in green. You must correct the eastern boundary of the viewshed by aligning it with the buffer. Note that, in addition to moving vertices, you have the ability to delete vertices.

5 Move your pointer over one of the vertices outside of the buffer, right-click, and select Delete Vertex.

The vertex has been removed, moving the boundary of the viewshed closer to the buffer.

6 Edit the remaining vertices falling outside the buffer by either deleting or moving them so that they snap to the buffer.

> **Note:** You can insert new vertices to give you additional control when reshaping a polygon. To add a vertex, scroll over the area where you want to add a vertex, right-click, and select Add Vertex.

7 Repeat steps 5 and 6 to correct the portion of the viewshed to the northwest of the camera.

8 On the Construction toolbar, click Finish to finalize the polygon edits.

9 On the Edit tab, click Save, and clear the selected feature.

3e

Exercise 3f: Edit polygon features (part 2)

Research highlight 3f: Merging CCTV viewsheds for focused police interventions

While CCTV video surveillance cameras are used by numerous police agencies to combat crime, specific surveillance barriers commonly prevent the full benefits of this technology from being realized (Piza et al. 2014b). Such surveillance barriers include the large numbers of CCTV cameras that operators are expected to monitor and the typically reactive manner by which police respond to incidents detected by CCTV. Given these limitations, the effect of stand-alone CCTV on crime has been limited. Conversely, systems that incorporate proactive monitoring strategies and deploy complementary interventions alongside CCTV have been shown to generate greater crime-control benefits than stand-alone camera deployment (Piza et al. 2019). In support of such focused strategies, it makes little sense for CCTV viewsheds lying directly adjacent to one another to be considered as separate entities. Rather, CCTV viewsheds that overlap or lie directly adjacent to one another should be merged for intervention purposes (Piza et al. 2015).

3f

In this exercise, you will work with the CCTV viewshed layer used in the prior exercise. You will make edits to the feature class by merging overlapping viewsheds into a single feature.

Note: If you closed the map after completing the prior exercise, open the Polygon, Edit map from the Catalog pane.

Export a new feature layer

1 Open the Merge Viewsheds 1 bookmark.

The map displays two overlapping viewsheds.

2 On the Map tab, click the Explore tool.

3 Click one of the overlapping viewsheds.

The pop-up window and map highlight confirm that the viewshed is not attached to the other; clicking the other overlapping viewshed does the same.

For intervention purposes, overlapping CCTV viewsheds in close proximity to one another need to be merged into single features. However, you do not want to merge the features in the original CCTV viewshed file. Rather, you are going to create a new feature layer where the merges will take place.

4 In the Contents pane, right-click Viewsheds, select Data, and select Export Features.

5 In the Export Features pane, for Output Name, type **Viewsheds_Intervention**, and click OK.

The next step is to select the two overlapping viewsheds. Before doing so, you want to prevent other layers from being inadvertently modified during the process.

6 In the Contents pane, right-click Viewsheds_Intervention, click Selection, and select Make This The Only Selectable Layer.

Merge overlapping features

1 In the Selection group on the ribbon, click the Select By Rectangle tool.

2 In the map view, draw a rectangle that encompasses the two overlapping viewsheds.

3 Deselect the tool once the rectangle is complete.

> **Note:** You can select additional viewsheds by pressing the Ctrl key while clicking the viewsheds individually.

4 In the Contents pane, right-click Viewsheds_Intervention, and click Attribute Table.

The two selected records are highlighted in the attribute table.

4	Polygon	Springfield Ave	& S. 18th St
5	Polygon	Springfield Ave	& S. 20th St

5 On the Edit tab, in the Features group, click Modify.

6 In the Modify Features pane, expand Construct, and click Merge.

The features that will be merged are identified in the Modify Features pane.

> **Note:** In the Merge panel, you can click Change The Selection to merge different features from those that are currently selected.

3f

7 Click Merge.

Selecting the features on the map and viewing the attribute table confirm that the merge was successful (in other words, there is only one feature where there were two).

> **Note:** The resulting features will be given the attributes of the features marked as Preserve in the Merge pane. If you want the attributes from a different feature to be preserved, drag that feature to the top of the list.

Update the new feature's attribute table

The Merge Features tool preserves the spatial aspects of each overlapping feature, combining them into a single geometry. However, the same does not happen for the attribute information. Rather than combining the attributes of each merged feature, ArcGIS Pro assigns the attributes of one of the features to the newly created feature. Often, the resulting attribute value will not accurately describe the merged feature. In the current example, the Camera field lists only one address (Springfield Ave. and S. 18th St.) despite the fact that the viewshed represents two separate addresses (Springfield Ave. and S. 18th St. and Springfield Ave. and S. 20th St.). You will conclude this exercise by updating this field.

1 In the attribute table, double-click the Camera field for the new feature.

2 In the text box, type **Springfield Ave. & S. 18th St.; Springfield Ave. & S. 20th St.,** and press Enter on the keyboard.

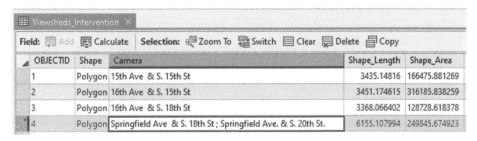

3 On the Edit tab, click Save.

To continue practicing merging features, you can open the Merge Viewsheds 2 bookmark and repeat all the steps in exercise 3f.

Summary

Chapter 3 built on the data processing foundation outlined in chapter 2. Chapter 2 provided skills about manipulating or querying existing data, whereas this chapter provided instruction creating and editing your own data. An increasing amount of data is available to crime analysts, but the need to create a new, custom dataset occurs frequently. Topics covered in chapter 3, including geocoding data, changing a dataset's projection, and creating and editing feature layers, are important skills for crime analysts.

Chapter 4
Maximizing attribute tables

Overview

In this chapter, you will learn how to create, edit, and manipulate data contained in attribute tables. Combined with spatial data, attribute tables allow analysts to perform in-depth analysis to support modern policing strategies.

You will acquire the following skills upon completing the exercises in this chapter:
- Enhancing feature attributes
- Creating summary tables
- Performing spatial joins
- Creating and calculating fields
- Performing table joins

Download and install the data

Before working on the exercises, you will obtain the necessary data.

1   Go to www.arcgis.com and sign in with your ArcGIS Online account credentials.

2   Type **Modern Policing Using ArcGIS Pro (Esri Press)** in the search box, and click the Groups tab. Make sure that the Only Search In Your Organization option is turned off.

3   Click the Modern Policing Using ArcGIS Pro (Esri Press) group.

4   On the group heading, click Content.

5   Click the chapter 4 file and download it.

6   Locate the zip file you downloaded to your local drive, right-click, and extract it to C:\ModernPolicing.

This will create a folder named Chapter4 in the ModernPolicing folder.

This folder contains an ArcGIS Pro project and data you will use for the exercises in this chapter.

Exercise 4a: Enhance feature attributes

Append attributes to points from polygons

1   Open Chapter4.aprx in C:\ModernPolicing\Chapter4.

    The project shows a map named Append Points. The map displays auto theft calls for service locations and police patrol districts in Jersey City, New Jersey. You will append attributes from the patrol districts layer to the auto theft points in this exercise.

2   In the Contents pane, open the attribute table for Auto Theft (JC).

    The auto theft layer currently contains nine attribute fields, with latitude and longitude coordinates to the far right of the table. You will add one attribute from the police districts layer to this table.

3   On the Crime Analysis Tab, in the Data Management group, click Enhance Attributes, and select Join Attributes From Polygon.

    The Join Attributes From Polygon tool opens in the Geoprocessing pane.

4   In the tool pane, for Target Point Features, select Auto Theft (JC).

5   For Input Polygon Features, select JCPD Districts.

6   Click the arrow to the right of Join Fields to open the fields list.

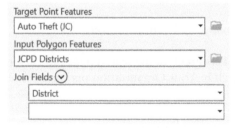

7   Click the check box next to District, click Add, and run the tool.

    The District attribute from JCPD Districts is added to the attribute table of Auto Theft (JC).

Summarize attribute values

You will create summary tables displaying the number of auto thefts within each police district in Jersey City.

1 In the Auto Theft (JC) attribute table, right-click the District field, and click Summarize.

2 In the Summary Statistics pane, type **District_Frequency** in the Output Table text box.

3 For Field, select District.

4 For Statistic Type, select Count, and click OK.

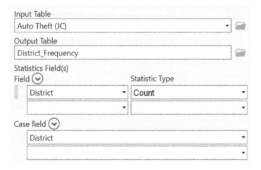

> **Note:** In this example, District was both the Case field and the Statistics field because you are interested in calculating the number of overall auto thefts within each district. You can select additional different case fields to calculate district counts across subsets of auto thefts (for example, patrol shifts) rather than the overall auto thefts.

District_Frequency appears beneath Standalone Tables in the Contents pane.

5 In the Contents pane, right-click District_Frequency, and click Open.

District	FREQUENCY
East	50
North	53
South	124
West	140

This summary table lists each of the District attribute values and the number of auto thefts in each district. The West district experienced the most auto thefts (140), followed by South (124), North (53), and East (50).

Extract a single date field to separate fields

For the final step in this exercise, you will extract information from a single date field to separate fields in the Auto Theft (JC) attribute table. The date and time of the event are currently contained within a single field in the attribute table (date). You will use the Crime Analysis tab to create new fields displaying the weekday, hour, month, day of month, and year of the event.

1   On the Crime Analysis tab, in the Data Management group, click Enhance Attributes, and select Add Date Attributes.

2   For Input Table, select Auto Theft (JC).

3   For Date Field, select Date.

4   Click Run.

Information from the date field is spread across five new fields in the Auto Theft (JC) attribute table: date_DW, date_HR, date_MO, date_DM, and date_YR.

tego	descriptio	shift	LAT	LONG	District	date_DW	date_HR	date_MO	date_DM	date_YR
rty Crimes	Motor Vehicle Theft	3	40.741535	-74.069939	West	Thursday	16	1	14	2016
rty Crimes	Motor Vehicle Theft	2	40.722004	-74.092201	West	Wednesday	8	5	18	2016
rty Crimes	Motor Vehicle Theft	3	40.739506	-74.052536	North	Wednesday	19	6	22	2016
rty Crimes	Motor Vehicle Theft	2	40.715771	-74.072502	West	Thursday	11	2	4	2016
rty Crimes	Motor Vehicle Theft	1	40.738319	-74.062447	West	Thursday	5	2	4	2016

Exercise 4b: Use spatial joins to summarize numeric attributes

Crime analysis highlight 4b: Using spatial joins to track crime changes in specialized areas

Police agencies collect and maintain large amounts of data, much of which is recorded at the address level. This spatial level is especially meaningful for individual cases and incidents, but it is also practical to assess the number of cases or events occurring within larger areas. Furthermore, there are additional incident attributes, such as officer response time and case disposition, that crime analysts need to account for in their reports. Spatial joins helped our agency examine such attributes of interest across defined boundaries that were useful to us and make informed decisions about how to allocate resources or respond to problems. We were able to quickly sum the number of calls for service, measure average officer response times, and determine the proportion of calls that resulted in an official police action throughout various areas of interest, such as our downtown district or public school zones. The data we gathered from performing spatial joins provided useful statistical outputs that could be shared interdepartmentally, at neighborhood watch meetings, or with interested city officials.
— *Nathan Connealy, crime analyst (former), Des Moines (Iowa) Police Department; doctoral candidate, John Jay College of Criminal Justice, City University of New York Graduate Center*

4b

Perform a spatial join

1   In the Catalog pane, in the Maps folder, double-click Spatial Join to open the map for this exercise.

The map displays surveillance camera schemes in the central area of Newark, New Jersey. These schemes consist of adjacent camera viewsheds merged into single features that were incorporated as patrol areas in the Newark Police Department's CCTV Directed Patrol Experiment (Piza et al. 2015). The map also contains a feature layer of crime incidents detected by Newark Police camera operators monitoring these cameras. In this exercise, you will perform a spatial join to measure the frequency and central tendency of the camera detections across the individual schemes.

Before performing the spatial join, you will familiarize yourself with the attributes in the Camera Detections layer.

2   Open the attribute table for the Camera Detections layer.

The attribute table contains information about the 475 crime incidents detected on these schemes since the inception of the Newark Police Department's video surveillance program. Several data points are available for each event. For this exercise, you will spatially join the crime points to the scheme polygons to measure the following within each feature: (1) the number of violent, drug, and disorder incidents detected by the camera operators, (2) the median queue minutes (time between crime detection and officer dispatch) and officer response time to each event, and (3) the number of enforcement actions resulting from the camera operator detections.

3 In the Contents pane, right-click CCTV Schemes, click Joins And Relates, and click Spatial Join.

4 In the tool pane, for Target Features, select CCTV Schemes.

5 For Join Features, select Camera Detections.

6 For Output Feature Class, type **Schemes_Detections**.

7 Click Fields beneath the text boxes.

You will now set the merge rules to generate the appropriate numeric attributes for the attribute fields of interest.

8 In the Output Fields section of the Spatial Join pane, click each attribute of interest and set the merge rules as follows:
- Que_Mins = Median
- Res_Mins = Median
- Violence = Sum
- Drugs = Sum
- Disorder = Sum
- Other = Sum
- All_Enforce = Sum

You will not preserve the remaining attributes in the polygon layer following the spatial join. You will remove these fields from the spatial join.

9   In the Output Fields section of the Spatial Join pane, click the following attributes, and click the red *X* to the right of each attribute to remove each field from the spatial join:
- Camera
- Shape_Length
- Shape_Area
- Time_Event
- Time_Dispatch
- Time_Arrive
- Tot_Mins
- Event
- Arrest
- Enforce_Other
- X
- Y

10   Click OK.

The Schemes_Detections layer created by the spatial join appears in the Contents pane. The attribute table includes the merged attributes from the Camera Detections file.

11   Open the attribute table for Schemes_Detections.

The merged attributes appear to the right of the attribute table. Note that any features with <Null> values did not intersect any points from the Camera Detections layer. The Join_Count field identifies the number of points that intersect the polygon. In the context of this example, the Join_Count field displays the number of crime incidents detected by camera operators in each scheme.

Summarize attributes by precinct

You will conclude this exercise by measuring how the merged attributes differ across the different police precincts represented in the camera schemes.

1   In the Schemes_Detections attribute table, right-click the Precinct field, and click Summarize.

2   In the Summary Statistics pane, for Output Table, type **Detections_Summary**.

3   In the Field section of the Summary Statistics pane, click each attribute of interest, and set the Statistic Type as follows:
    - Join_Count = Sum
    - Que_Mins = Median
    - Res_Mins = Median
    - Violence = Sum
    - Drugs = Sum
    - Disorder = Sum
    - Other = Sum
    - All_Enforce = Sum

4   Ensure that Case Field is set to Precinct.

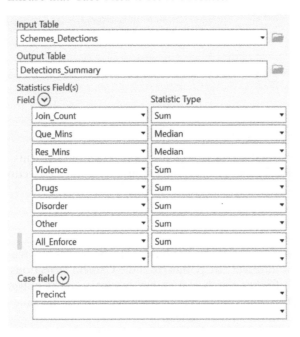

5   Click OK.

6   In the Contents pane, right-click Detections_Summary, and click Open.

This summary table lists the attribute values across police precincts. For example, the camera schemes in the third precinct detected 178 crime incidents, the majority of which (101) were disorder events. The median officer response time to these incidents was 7.075 minutes.

OBJECTID *	Precinct	FREQUENCY	SUM_Join_Count	MEDIAN_Que_Mins	MEDIAN_Res_Mins	SUM_Violence	SUM_Drugs	SUM_Disorder	SUM_Oth
1	<Null>	9	0	<Null>	<Null>	<Null>	<Null>	<Null>	<Nul
2	2nd	6	50	9.11	7.46	9	14	22	
3	3rd	6	178	10.63	7.075	4	51	101	
4	4th	7	113	22.38	9.7	10	42	45	
5	5th	10	134	6.9	9.51	10	51	56	

Exercise 4c: Perform table joins

In this exercise, you will combine four separate attribute tables within a single layer through a series of table joins. It is helpful to inspect the attribute table of the primary layer before performing table joins to determine how the disparate tables will be combined.

Join feature classes with an identical number of features

1   In the Catalog pane, in the Maps folder, click Table Join to open the map for this exercise.

The map displays all counties across the states of Connecticut, New Jersey, and New York. The Contents pane contains two additional layers with their visibility turned off: Counties (CT, NJ, NY) – Violent Crime, and Counties (NJ) – Other Deaths. A stand-alone table (Counties_NJ_Economics) also appears at the bottom of the Contents pane.

2   Open the attribute table for Counties (CT, NJ, NY).

Five fields contain identifying information for each of the 91 counties included in the feature class: FIPS, County Name, State Abbreviation, State Name, and County. The FIPS field is the unique identifier that will be used to join the tables.

3   Open the attribute table for Counties (CT, NJ, NY) – Violent Crime.

The attribute table for this feature class includes the number of violent crimes, violent crime rate, firearm fatalities, and firearm fatality rate for all 91 counties in Connecticut, New Jersey, and New York. You will join this to the main counties layer using the FIPS field.

4   In the Contents pane, right-click Counties (CT, NJ, NY), click Joins And Relates, and click Add Join.

5   In the Add Join tool pane, for Input Join Field, select FIPS.

6   For Join Table, select Counties (CT, NJ, NY) – Violent Crime.

7   For Join Table Field, select FIPS.

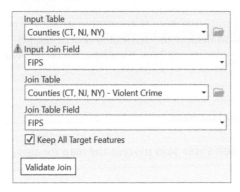

8   Click OK.

The Counties (CT, NJ, NY) attribute table now contains the attributes from the violent crime layer. FIPS appears twice, indicating that attributes from two separate tables are now joined. The violent crime information from Counties (CT, NJ, NY) – Violent Crime now appears alongside the county identifier in Counties (CT, NJ, NY).

reviation	State Name	County	Shape_Length	Shape_Area	OBJECTID	FIPS	Violent Crimes	Violent Crime Rate	Fire
	Connecticut	Fairfield	345553.245214	2995955253.5	1	09001	2428	264.01	
	Connecticut	Hartford	336251.930587	3501191837.5	2	09003	2565	293.17	
	Connecticut	Litchfield	309330.853084	4404300155	3	09005	129	110.46	
	Connecticut	Middlesex	226483.058917	1780751786	4	09007	123	114.34	
	Connecticut	New Haven	338389.22446	2861365642	5	09009	3084	381.33	

The attribute table is currently set to display an alias for the field name. This alias is a flexible and more readable option that does not need to follow the conventions required by the database structure.

9 In the upper right of the attribute table, click the More Options button (three lines) to open the options menu, and uncheck Show Field Aliases to turn off the alias display.

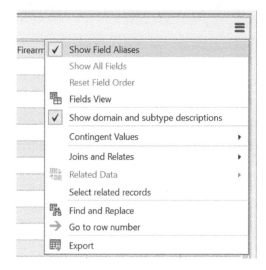

10 Scrolling through the attribute table, expand the field margins to see the full field names.

The two fields that both used the alias FIPS are now displayed as Counties_CT, NJ, NY.GEOID10 and Counties_CT_NJ_NY_ViolentCrime.GEOID10. This view can be helpful in identifying where a field comes from while you work with joined tables.

Join feature classes with different numbers of features

In the prior example, each feature in the input table had a unique match in the join table. You can also perform table joins when the tables contain different numbers of features. This will be demonstrated by joining tables containing information for only the 21 counties in New Jersey to the Counties (CT, NJ, NY) layer.

1 In the Contents pane, right-click Counties (CT, NJ, NY), click Joins And Relates, and click Add Join.

In the tool pane, the attribute values in the Input Join Field list look different now that the layer has an active table join. The name of the layer is shown first, followed by the name of the field (for example, Counties (CT, NJ, NY) or Counties (CT, NJ, NY) – Violent Crime). Despite the different look, you will still use the GEOID10 attribute as the join field.

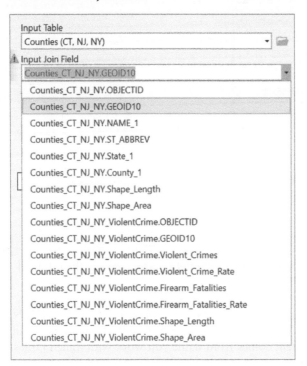

2   In the Add Join tool pane, for Input Join Field, select Counties_CT_NJ_NY.GEOID10.

3   For Join Table, select Counties (NJ) – Other Deaths.

4 For Join Table Field, select FIPS.

Input Table
Counties (CT, NJ, NY)

⚠ Input Join Field
Counties_CT_NJ_NY.GEOID10

Join Table
Counties (NJ) - Other Deaths

Join Table Field
FIPS

☑ Keep All Target Features

Validate Join

5 Click OK.

The unique identifiers for Counties (NJ) – Other Deaths are contained in the FIPS field, which is still using the alias for the field name. This data is identical to the GEOID10 field in the other tables.

The Counties (CT, NJ, NY) attribute table is now joined with Counties (NJ) – Other Deaths. All 91 counties appear in the layer, but all counties outside of New Jersey have <Null> values listed for their attributes.

You will now join one final New Jersey table to Counties (CT, NJ, NY). You will remove all features from Counties (CT, NJ, NY) without a match in the join table, leaving only New Jersey Counties visible in the map view and attribute table.

6 In the Contents pane, under Standalone Tables, open the Counties_NJ_Economics attribute table.

This attribute table includes various economic indicators for the 21 counties in New Jersey. Again, the unique identifiers for the counties are contained in the FIPS field.

7 In the Contents pane, right-click Counties (CT, NJ, NY), click Joins And Relates, and click Add Join.

8 In the tool pane, for Input Join Field, select Counties_CT_NJ_NY.GEOID10.

9 For Join Table, select Counties_NJ_Economics.

10 For Join Table Field, select FIPS.

4c

11 Uncheck Keep All Target Features.

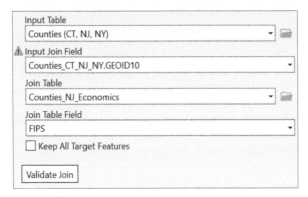

12 Click OK.

The Counties (CT, NJ, NY) attribute table now includes the economic indicators contained within Counties_NJ_Economics. However, only the 21 counties in New Jersey present in both the input and the join tables are visible.

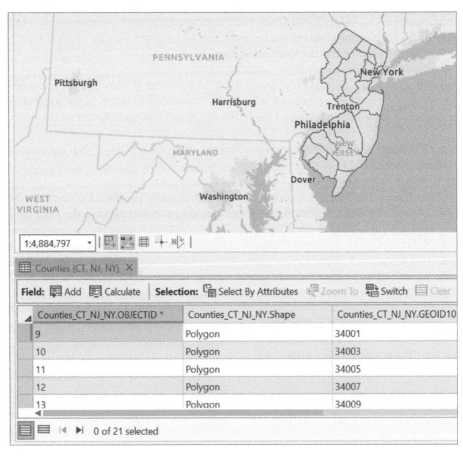

Export joined tables to a new feature class

It is important to note that table joins are temporary connections between multiple tables in your ArcGIS Pro project. The joins do not alter the structure of your files. As such, the table joins are only visible within the project in which they were created. You must export the joined layer as a stand-alone feature class to make the full set of attribute fields visible outside the current table join.

1   In the Contents pane, right-click Counties (CT, NJ, NY), click Data, and click Export Features.

2   In the Export Features tool pane, for Output Name, type **NJ_Counties_All**.

3   Click OK.

NJ_Counties_All now appears in the Contents pane. This layer contains all the attributes that were included in the four tables that you previously joined. You will remove the table joins from Counties (CT, NJ, NY).

4   In the Contents pane, right-click Counties (CT, NJ, NY), click Joins And Relates, and click Remove All Joins. Click Yes to confirm.

The Counties (CT, NJ, NY) attribute table now includes only the fields from the original attribute table.

4c

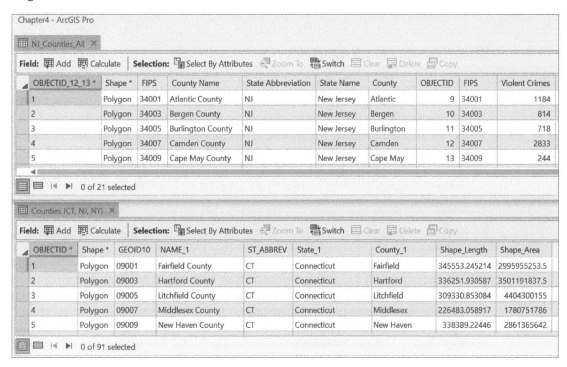

Exercise 4d: Create and calculate fields

Research highlight 4d: Calculating location quotients

Measuring crime in terms of the overall population at risk of victimization is an inherent challenge for crime analysts. Crime rates, which control for the number of people in the population (for example, crime per 100,000 residents), likely represent the most common approach. However, rates can be problematic, as people are typically outside the neighborhood they live in for large portions of the day (Brantingham & Brantingham 1998). Certain neighborhoods often have high crime counts but a low residential population (for example, downtown commercial areas), resulting in exaggerated crime rates (Andresen et al. 2009). Given the challenges associated with crime rates, researchers have begun to use location quotients to contextualize crime levels. The location quotient represents the percentage of a particular criminal activity within a given area relative to the percentage of that same criminal activity in the entire study area (Andresen et al. 2009). Location quotients have recently been adapted to overcome challenges present when units of analysis are not uniform in size (Ratcliffe 2010). By ignoring the different sizes of spatial units, analysts may generate misleading maps suggesting that crime is higher in one area than another without controlling for unequal size of geographies (Harries 1999). The location quotient is calculated through the following formula:

$$LQ = (c_i/a_i)/(c_R/a_R)$$

In this formula, *c* is the frequency of crime and *a* is the area of a subset location (*i*) of a larger region (*R*) (Ratcliffe 2010, 18). Research on CCTV has used the location quotient to measure crime given the differing geographic sizes of camera viewsheds (Piza et al. 2014a).

Manually create and calculate fields

In this exercise, you will calculate two location quotients for each viewshed. The first location quotient will measure the crime rate within the viewshed compared with the surrounding police precinct during the one-year pre-camera installation period. The second location quotient will measure the crime rate within the viewshed compared with the surrounding police precinct during the one-year post-camera installation period. Before calculating the location quotients, you will sum the four crime types together into an "all crime" category by creating and calculating fields in the attribute table.

1   In the Catalog pane, in the Maps folder, click Create Fields to open the map for this exercise.

The map shows all 117 camera viewsheds in Newark, New Jersey. The attribute table contains crime counts for each viewshed as well as its encompassing police precinct, for both the one-year pre- and post-camera installation periods. Four total crime types are represented in the table: auto theft (Auto_), theft from auto (TFA_), robbery (Rob_), and gun violence (murders and nonfatal shootings, Gun_). The square footage of the viewsheds (Total_SqFt) and surrounding precinct (P_SqFt) are also included.

2 In the Camera Viewsheds Crime attribute table, click the Add Field button.

You will add six fields to the attribute table. You will calculate values in these fields throughout the remainder of this exercise.

3 For each new field, type the name in the Field Name and Alias columns, and select the appropriate Data Type from the list, as follows:
- **All_pre** (Data Type = Long)
- **All_pst** (Data Type = Long)
- **P_All_pre** (Data Type = Long)
- **P_All_pst** (Data Type = Long)
- **LQ_All_pre** (Data Type = Double)
- **LQ_All_pst** (Data Type = Double)

✓	☐	All_pre	All_pre	Long
✓	☐	All_pst	All_pst	Long
✓	☐	P_All_pre	P_All_pre	Long
✓	☐	P_All_pst	P_All_pst	Long
✓	☐	LQ_All_pre	LQ_All_pre	Double
✓	☐	LQ_All_pst	LQ_All_pst	Double

4 On the Fields tab on the ribbon, click Save.

5 Close the Create Fields table.

6 In the Camera Viewsheds Crime attribute table, scroll to the right.

The table contains the six new fields you just created.

7 In the field headings, right-click All_pre, and click Calculate Field.

You will sum each of the pre-crime categories to measure the total crime events occurring during the one-year pre-camera installation period. (**Hint:** In the Calculate Field pane, under Fields, double-click a field name to add it to the expression. Under Helpers, click the necessary mathematical symbols [for example, +] to add them to the expression box.)

8 In the All_pre expression text box, build this expression: **!Auto_pre! + !TFA_pre! + !Rob_pre! + !Gun_pre!**

9 Click OK.

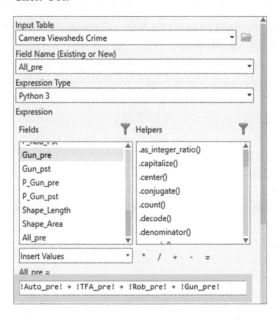

10 Repeat steps 7 and 8 for the remaining three new attribute fields with Long data types (All_pst, P_All_pre, and P_All_pst), building expressions as follows:

- All_pst = **!Auto_pst! + !TFA_pst! + !Rob_pst! + !Gun_pst!**
- P_All_pre = **!P_Auto_pre! + !P_TFA_pre! + !P_Rob_pre! + !P_Gun_pre!**
- P_All_pst = **!P_Auto_pst! + !P_TFA_pst! + !P_Rob_pst! + !P_Gun_pst!**

Each feature now has values for the four Long attributes.

All_pre	All_pst	P_All_pre	P_All_pst
5	11	1190	943
6	9	2183	1962
11	9	2183	1962
2	3	2183	1962
16	15	2183	1962

Save an expression

ArcGIS Pro allows users to save and load expressions into the Calculate Field tool pane. This can be helpful when writing complex expressions that you will use again later. You will save and load expressions that calculate the location quotients for the pre- and post-crime levels in the viewsheds.

1 In the Camera Viewsheds Crime attribute table, right-click the LQ_All_pre field, and click Calculate Field.

2 In the LQ_All_pre expression text box, build this expression: (**!All_pre! / !Total_SqFt!) / (!P_All_pre! / !P_SqFt!)**

3 At the bottom of the Calculate Field pane, click the green arrow to export the expression.

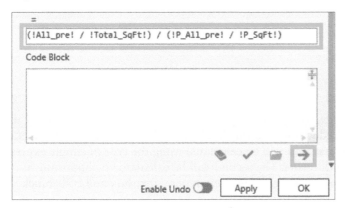

4 Browse to C:\ModernPolicing\Chapter4\Expressions.

5 In the Name text box, type **LQ_All_pre**, and click Save.

The location quotient expression for pre-camera installation crime is now saved for future use.

6 In the Calculate Field pane, click OK to calculate the pre-camera installation location quotient.

Load an expression

You will use a previously created expression to calculate the post-camera installation location quotient.

1 In the Camera Viewsheds Crime attribute table, right-click the LQ_All_pst field, and click Calculate Field.

2 At the bottom of the Calculate Field pane, click the Import button (folder icon) to import an expression.

3 Browse to C:\ModernPolicing\Chapter4\Expressions.

4 Click ALL_LQ_Pst.cal.

5   Click OK.

```
=
(!All_pst! / !Total_SqFt!) / ( !P_All_pst! / !P_SqFt!)
```

The location quotient expression is now in the expression text box.

6   Click OK.

Calculate fields using the code block

You will conclude this exercise by creating and calculating two fields to identify the cameras that generated reductions in crime levels from the pre- to the post-camera installation period. The first field will be calculated using the type of simple expression that you worked with earlier. The second will be calculated by importing a Python script into the Calculate Field pane's expression text box and code block.

1   In the Camera Viewsheds Crime attribute table, click the Add Field button.

2   For each new field, type the name in the Field Name and Alias columns, and select the appropriate data type from the list, as follows:
    - **LQ_diff** (Data Type = Double)
    - **LQ_neg** (Data Type = Short)

3   On the Fields tab on the ribbon, click Save.

4   Close the Create Fields table.

5   In the Camera Viewsheds Crime field, right-click LQ_diff, and click Calculate Field.

6   In the Calculate Field pane, in the LQ_Diff expression text box, type this expression: **!LQ_All_pst! - !LQ_All_pre!.**

7   Click OK.

The LQ_diff field reflects the change in crime location quotients from the pre- to the post-camera installation period. Camera viewsheds with positive LQ_diff values experienced a crime level increase, while cameras with negative values experienced a reduction. You will calculate the LQ_neg field so that the viewsheds experiencing crime reductions can be readily identified in the attribute table.

8   In the Camera Viewsheds Crime table, right-click LQ_neg, and click Calculate Field.

9   At the bottom of the Calculate Field pane, click the Import button to import an expression.

10 In the Chapter4\Expressions folder, double-click Neg_calc.cal.

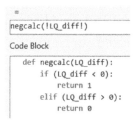

```
=
negcalc(!LQ_diff!)
```

Code Block

```
def negcalc(LQ_diff):
    if (LQ_diff < 0):
        return 1
    elif (LQ_diff > 0):
        return 0
```

A custom Python script is added to the Calculate Field pane. The expression text box contains the name of the Python script rather than a value that should be assigned to the field. The code block defines the parameters of the Python script. In this example, the Python script assigns a value of 1 to the LQ_neg field if LQ_diff is less than (<) 0 and a value of 0 if LQ_diff is greater than (>) 0.

11 Click OK.

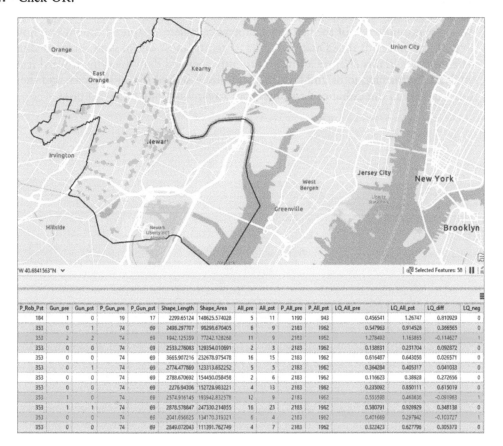

P_Rob_Pst	Gun_pre	Gun_pst	P_Gun_pre	P_Gun_pst	Shape_Length	Shape_Area	All_pre	All_pst	P_All_pre	P_All_pst	LQ_All_pre	LQ_All_pst	LQ_diff	LQ_neg
184	1	0	19	17	2299.65124	148625.574028	5	11	1190	943	0.456541	1.26747	0.810929	0
353	0	1	74	69	2498.297707	98298.670405	8	9	2183	1962	0.547963	0.914528	0.366565	0
353	2	2	74	69	1942.125359	77242.128268	11	9	2183	1962	1.278492	1.163865	-0.114627	1
353	0	0	74	69	2533.276083	129354.010691	2	3	2183	1962	0.138831	0.231704	0.092872	0
353	0	0	74	69	3665.907216	232678.975478	16	15	2183	1962	0.616487	0.643058	0.026571	0
353	0	1	74	69	2774.477869	123313.652252	5	5	2183	1962	0.364284	0.405317	0.041033	0
353	0	0	74	69	2788.670692	154450.058458	2	6	2183	1962	0.116623	0.38928	0.272656	0
353	0	0	74	69	2276.94306	152728.983221	4	13	2183	1962	0.233092	0.850111	0.615019	0
353	1	0	74	69	2574.916145	193942.832578	12	9	2183	1962	0.555598	0.463636	-0.091963	1
353	1	1	74	69	2878.578847	247330.214855	16	23	2183	1962	0.380791	0.928929	0.348138	0
353	0	0	74	69	2041.856825	134170.319321	6	4	2183	1962	0.401669	0.297942	-0.103727	1
353	0	0	74	69	2849.072043	111391.762749	4	7	2183	1962	0.322423	0.627796	0.305373	0

4d

- The LQ_neg field is now populated and can be used to quickly identify camera viewsheds with decreasing crime levels. For example, running a Select By Attribute query with an expression that states **Where LQ_neg Is Equal To 1** would select the relevant viewsheds in both the map view and the attribute table.

> **Note:** The code block allows users to create complex expressions. While this exercise used Python, the code block also supports Arcade and SQL coding expressions. For more information on the code block and using scripting languages in ArcGIS Pro, search ArcGIS Pro documentation (pro.arcgis.com) for these topics:
> - Tool Reference > Geoprocessing Tools > Data Management Toolbox > Fields Toolset > Calculate Field (Data Management)
> - Tool Reference > Geoprocessing Tools > Data Management Toolbox > Fields Toolset > Fields Toolset Concepts > Calculate Field Python Examples

Summary

With the groundwork of the last two chapters complete, chapter 4 taught you how to perform more advanced analysis tasks to get the most out of your attribute data. The exercises in this chapter allow you to move beyond data querying toward true analysis. Performing joins and creating fields are essential for all analysts, and the functionality in the Enhance Attributes tool streamlines much-needed tasks.

Chapter 5
Identifying crime hot spots and tracking crime in target areas

Overview

In this chapter, you will learn how to use the various tools for identifying hot spots and measuring crime changes in areas of interest. The different techniques covered in this chapter can inform a variety of crime analysis tasks, including target area identification and program evaluation. Many of these tools are new to the ArcGIS Pro crime analysis solution.

You will acquire the following skills upon completing the exercises in this chapter:
- Creating crime density maps
- Measuring crime density change over time
- Performing 80-20 analysis
- Creating optimized hot spots
- Creating street intersection points and Thiessen polygons
- Counting incidents and calculating percent changes in polygons

Download and install the data

Before working on the exercises, you will obtain the necessary data.

1   Go to www.arcgis.com and sign in with your ArcGIS Online account credentials.

2   Type **Modern Policing Using ArcGIS Pro (Esri Press)** in the search box, and click the Groups tab. Make sure that the Only Search In Your Organization option is turned off.

3   Click the Modern Policing Using ArcGIS Pro (Esri Press) group.

4   On the group heading, click Content.

5   Click the chapter 5 file and download it.

6   Locate the zip file you downloaded to your local drive, right-click, and extract it to C:\ModernPolicing.

This will create a folder named Chapter5 in the ModernPolicing folder.

This folder contains an ArcGIS Pro project and data you will use for the exercises in this chapter.

> **Research highlight 5: Applying multiple tools in hot spot identification**
>
> The identification of crime hot spots has become a primary function in modern policing. Police can more readily focus crime prevention resources after identifying where crime concentrates. Hot spots policing has a demonstrated record of success (Braga et al. 2019) and offers a more efficient method of policing than an exclusive focus on individual offenders (Lum & Koper 2017; Lum et al. 2011; Weisburd 2011, 2015). The renewed emphasis on microplaces in policing has also granted crime analysts a more central role in crime prevention efforts (Piza 2019). However, too often lost in the discussion are the technical processes to identify geographic crime hot spots. As argued by Johnson et al. (2015), police agencies interested in conducting hot spots policing need to know much more than that the practice works; they need to know what density of crime defines a hot spot, how hot spots should be selected for intervention, and the number of hot spots that officers can realistically address during a given shift. Adding to this complexity, multiple hot spot identification techniques exist, each of which may be applicable to different types of crime, situational context, and police intervention priorities (Haberman 2017). Police stand to benefit from incorporating a range of hot spot identification strategies in their crime analysis functions.

Exercise 5a: Create crime density maps and measure density change

In this exercise, you will create a series of density maps using this data. Also commonly referred to as *hot spot maps*, density maps provide a convenient way to display large amounts of spatial data in an easy-to-understand format. As opposed to looking at hundreds or even thousands of data points, a density map converts points into a visual representation similar to that of a weather map. ArcGIS Pro has added unique functionality to density maps through the Density Change tool. The Density Change tool combines two density maps to show how the concentration of points has changed over time.

Create kernel density maps

1   Open Chapter5.aprx in C:\ModernPolicing\Chapter5.

The project shows a map named Density. This map contains two layers displaying shootings in Rochester, New York: one for the year 2019 and one for 2018.

2   On the Crime Analysis tab, in the Analysis Tools Gallery, click Kernel Density.

3   In the Kernel Density tool pane, for Input Point Or Polyline Features, select Shootings_2019.

4 For Output Raster, type **Density_2019**.

> **Note:** File names for rasters must be fewer than 13 characters.

5 For Output Cell Size, type **250**.

6 For Search Radius, type **1000**.

> **Note:** Opinions vary about what values should be used for cell size and search radius. Familiarity with your data is a good place to start. Ideally, cell size should be something meaningful and appropriate to your study area; examples include a fraction of a standard city block (in other words, the 250 value for the output cell size) is roughly half the size of most blocks in Rochester). Similarly, your search radius should make sense to your data. The value of 1,000 used for the search radius means that the tool will search for data points within roughly two blocks of one another. Values that are too small or too large for your data may provide results that have limited practical value. In the end, experimenting with the tool parameters can help you discover results that are valuable to your audience.

7 For Area Units, select Square Feet.

5a

> **Note:** The format of the cell size and search radius is contingent on the projection of your data. The Rochester shooting data is projected in the State Plane New York West coordinate system, which enables users to choose the area units. Other projections, such as WGS84 Web Mercator, do not allow for customized area units. You also need to be mindful of the cell size and radius value interpretations across projections. Typing 1000 (decimal degrees) with WGS84 Web Mercator is much different from typing 1000 (feet) with State Plane New York West.

8   For Input Barrier Features, select Rochester_Border.

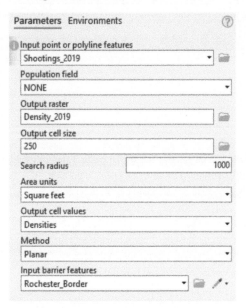

Barrier features limit the search area of raster processes to areas that can potentially experience a crime (in this example, the Rochester city limits).

9   Click Run.

The Density_2019 layer now appears in the Contents pane.

10 Turn off the Shootings_2019 and the Shootings_2018 layers.

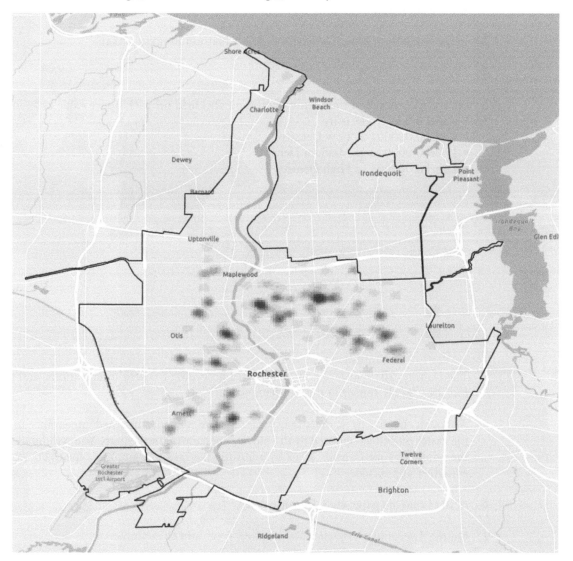

11 Repeat the previous steps to create a kernel density layer named Density_2018 from the Shootings_2018 feature class.

Symbolize the density layers

The density layers appear in the Contents pane with a default symbology. You will customize the symbology of both layers.

1 In the Contents pane, click Density_2019.

2 On the Appearance tab, click Symbology.

The Symbology tool opens. Similar options offered with vector layers are available with raster maps.

3 In the Symbology pane, for Method, select Natural Breaks (Jenks).

4 For Classes, select 4.

5 On the Classes tab at the bottom, edit the Label field for each upper value as follows:
  - Label 0.0 as **No Density**
  - Label 0.000001 as **Low Density**
  - Label 0.000002 as **Medium Density**
  - Label 0.000005 as **High Density**

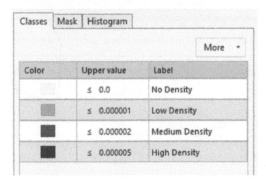

The Density_2019 layer is now symbolized according to these specifications, and the four new labels appear in the Contents pane.

Although effective as is, users may prefer a more refined, smoothed appearance to the raster output as opposed to the blocky appearance in the map. You will finalize the symbology for Density_2019 by adjusting the raster resampling type to create a smoother appearance.

6 In the Contents pane, click Density_2019.

7 On the Appearance tab, click Resampling Type, and click Cubic.

The raster now has a smoother look.

8 Repeat the prior steps to symbolize Density_2018.

9 Select a different color ramp to distinguish Density_2018 from Density_2019.

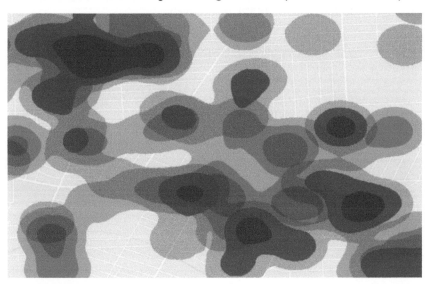

Create a density change map

The density maps you just created are helpful but do not provide context about how hot spots may have changed over time. You will complete this exercise by creating a density change map that measures the movement of shooting hot spots from 2018 to 2019.

1 On the Crime Analysis tab, in the Analysis Tools Gallery, click Minus.

2 In the tool pane, for Input Raster Or Constant Value 1, select Density_2019.

3 For Input Raster Or Constant Value 2, select Density_2018.

4 For Output Raster, type **Density_Change**, and run the tool.

Density change maps appear in the Contents pane in stretch format by default, which is difficult to interpret. You will conclude this exercise by symbolizing the density change layer using the Classify method.

5 On the Appearance tab on the ribbon, click Symbology, and click Classify.

6 In the Symbology pane, for Classes, select 3.

7 At the bottom of the pane, click the Histogram tab.

The Histogram tab provides a visual depiction of the density value range in Density_Change. You will manually adjust the values so that the upper value of the middle range (in orange) equals –0.000001 and the lower value equals 0.00000.

8 Drag the arrows next to the upper and lower two middle values in the histogram until they equal –0.000001 and 0.00000, respectively.

> **Note:** When you move the arrows, the symbology method automatically changes to Manual Interval.

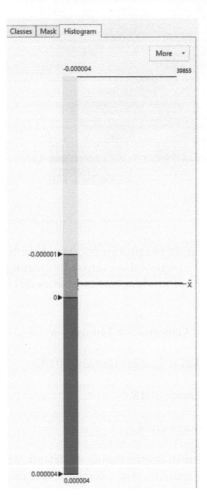

9 In the Symbology pane, click the Classes tab.

The Classes tab shows the values and labels for the three density categories.

10 For the second class, click the polygon in the Color column, and change the color to No Color.

This reflects that all raster cells in this category experienced no change in density value from 2018 to 2019.

11 In the Symbology pane, edit the Label field for each value as follows:
   - Replace <= –0.000001 with **Decrease**
   - Replace <= 0.0 with **No Change**
   - Replace <= 0.000004 with **Increase**

12 Turn off the Density_2018 and Density_2019 layers.

13 Perform a cubic resampling, as you did for the Density_2018 and the Density_2019 layers.

Density_Change is now symbolized according to the change in shooting concentration from 2018 to 2019.

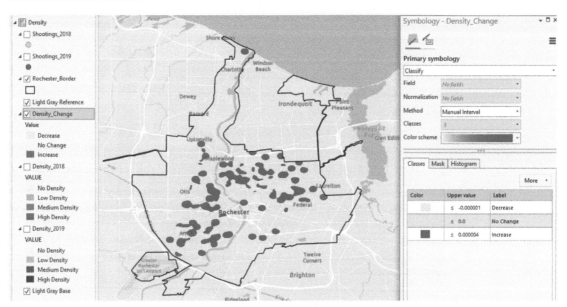

5b

Exercise 5b: Perform 80-20 analysis

You will use the 80-20 Analysis tool in the following exercise to identify how shootings in Rochester, New York, are distributed across individual addresses. Based on the Pareto Principle, the 80-20 Analysis tool creates a graduated symbol layer for points with multiple data records at the same location.

Use the 80-20 Analysis tool

1 In the Catalog pane, in the Maps folder, click 80-20 to open the map for this exercise.

The map displays all shooting locations in Rochester, New York, over the five-year period from 2015 to 2019.

2 In the tool pane, on the Crime Analysis tab, in the Analysis Tools Gallery, click 80-20 Analysis.

The 80-20 Analysis tool pane appears.

3 For Input Point Features, select RPDShootings2015_2019.

4 For Output Feature Class, type **RPDShootings2015_2019_8020Analysis.**

5 For Output Fields, select Address to preserve it in the attribute table of the output.

Input Point Features
RPDShootings2015_2019
Output Feature Class
RPDShootings2015_2019_8020Analysis
Cluster Tolerance

Output Fields ⌄
Address

6 Keep Cluster Tolerance and all other tool parameters blank, and run the tool.

> **Note:** The 80-20 Analysis tool considers any points falling within the cluster distance as part of the same cluster. When the default values are used, each individual address is a self-contained cluster. When a cluster distance is specified, the resulting points represent the central point of the cluster. It is important to note that the resulting point may be in a place where no actual crime events took place when a cluster distance is specified.

The new 80-20 analysis layer now appears in the Contents pane.

7 Turn off the RPDShootings2015-1019 layer.

The 80-20 Analysis tool collapsed the overall shooting points into individual points representing each shooting incident address. Points are symbolized according to the number of incidents occurring at the address. You will zoom in to the addresses experiencing the most shooting incidents.

8 Open the attribute table for the RPDShootings2015_2019_8020Analysis layer.

The attribute table shows that the 902 shooting events in Rochester were collapsed into 723 clusters, according to the parameters added to the 80-20 Analysis tool. The attribute table lists the count of points (in this example, shootings) occurring at each address and the percentage of overall shootings the address accounts for. The Cumulative Percentage field can be used to determine how many overall events could potentially be addressed by identifying a subset of clusters. In this example, the top

20 addresses accounted for nearly 10 percent of shootings in Rochester from 2015 through 2019. The top two addresses accounted for eight shootings each.

9 In the attribute table, select the first two rows.

10 Right-click, and click Zoom To.

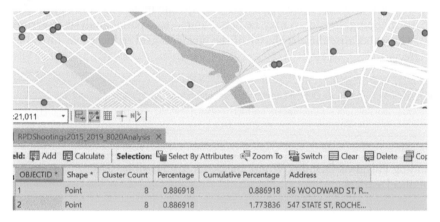

The map now displays the two addresses with the most shootings during the analysis period.

Exercise 5c: Create optimized hot spots

In this exercise, you will use the Optimized Hot Spot Analysis tool to create a map of statistically significant hot and cold spots throughout Chicago. The simultaneous display of high values (hot spots) and low values (cold spots) distinguishes the Optimized Hot Spot Analysis tool from other hot spot detection techniques. Such knowledge can inform police leadership whether problem-solving projects are effective at reducing a given problem or whether resources need to be adjusted and deployed to a different area where the problem is emerging.

Optimize hot spots within a hexagon grid

1 In the Catalog pane, in the Maps folder, double-click Optimized to open the map for this exercise.

The map displays the locations of all violent crime incidents in Chicago, Illinois, during 2014, as well as the boundaries of the 25 police districts in the city. A stand-alone table named violence_harmscore is also included.

2 On the Crime Analysis tab, in the Analysis Tools gallery, click Optimized Hotspot.

3 In the tool pane, for Input Features, select Chicago Violent Crime.

5c

4 For Output Features, type **ViolentCrime_Optimized.**

5 For Incident Data Aggregation Method, select Count Incidents Within Hexagon Grid.

The default is to use a fishnet of equally sized square-shaped grids as the point aggregation method. While square grids are perhaps the most common approach for thematic mapping, hexagons may be better suited for spatial analysis in certain instances. This is because the circularity of the hexagon grid reduces bias introduced by edge effects.

> **Note:** For more information about the benefits of hexagon grids in point pattern analysis, see ArcGIS Pro > Tool Reference > Geoprocessing Tools > Spatial Statistics Toolbox > Why Hexagons?

6 For Bounding Polygons Defining Where Incidents Are Possible, select ChicagoPoliceDistricts.

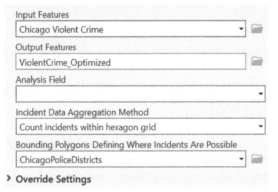

While not required, specifying a bounding polygon refines the hot spot analysis by limiting the search area. This allows the tool to distinguish between areas that truly did not experience crime from areas for which crime data is merely unavailable because the area falls outside the current study setting (similar to the barrier features used in density map creation).

7   Click Run.

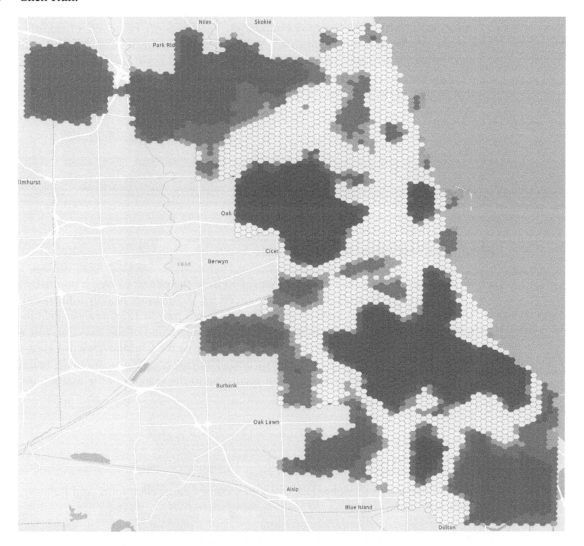

The newly created ViolentCrime_Optimized layer appears in the Contents pane. The layer shows where violent crime points create statistically significant hot spots (the red clusters) and cold spots (the blue clusters).

> **Note:** Specifying ChicagoPoliceDistricts as the bounding polygon enables the tool to more readily identify statistically significant cold spots. Running the Optimized Hot Spot Analysis tool without a bounding polygon would result in certain significant cold spots (for example, the vicinity of O'Hare International Airport in the northwest area of the city) being identified as not significant. To examine this result, repeat the prior steps while keeping the bounding polygons text box empty to see the difference, saving the output as ViolentCrime_NoBounding.

The attribute table for the VioletCrime_Optimized layer provides information about each of the hexagon grids. The Counts field displays the number of violent crime points within the hexagon boundaries. The GiZScore Fixed 3974 field lists the z-score calculated from the mean and standard deviation of the Counts value while accounting for a feature's total number of nearest neighbors. The GiPValue Fixed 3974 field lists the corresponding confidence level used to determine the statistical significance of the hot spot (in other words, positive z-score) or cold spot (negative z-score).

OBJECTID *	Shape *	SOURCE_ID	Counts	Shape_Length	Shape_Area	GiZScore Fixed 3974	GiPValue Fixed 3974
1	Polygon	1	13	4763.139355	1637329.027699	1.928043	0.05385
2	Polygon	2	1	4763.139683	1637329.253256	1.294184	0.195602
3	Polygon	3	0	4763.139027	1637328.80214	-2.427557	0.015201
4	Polygon	4	0	4763.139683	1637329.253256	-1.840713	0.065664
5	Polygon	5	1	4763.139683	1637329.253256	-1.71878	0.085654

Optimize hot spots using an analysis field

When creating optimized hot spot maps, you have the option to weigh points by a numeric field in the attribute table so that certain points more heavily influence the resulting hot spot map. Such an attribute field is referred to as an analysis field in the Optimized Hot Spot Analysis tool. For the next portion of this exercise, you will first join the violence_harmscore stand-alone table to the Chicago Violent Crime feature layer to optimize by an analysis field. The violence_harmscore table assigns weights to each violent crime incident according to the severity of the weapon used, as follows:

- Firearm = 10
- Knife = 5
- Other weapon = 3
- No weapons used = 1

1   In the Contents pane, right-click Chicago Violent Crime, click Joins And Relates, and click Add Join.

2   In the Add Join pane, for Input Join Field, select ID.

3   For Join Table, select violence_harmscore.

4   For Join Table Field, select ID.

5   Run the tool.

The attribute fields from violence_harmscore are added to the Chicago Violent Crime attribute table.

You will export this result to a new feature class.

6   Right-click Chicago Violent Crime, click Data, and click Export Features.

7   In the Export Features tool pane, for Output Name, type **ChicagoViolentCrime_ Harmscore.**

8   Run the tool.

You will now use the Harmscore field to create an optimized hot spot map.

9   On the Crime Analysis tab, in the Analysis Tools Gallery, click Optimized Hotspot.

10   In the tool pane, for Input Features, select Chicago ViolentCrime_Harmscore.

11   For Output Features, type **ViolentCrime_Harm.**

12   For Analysis Field, select Harmscore.

13   Run the tool.

The newly created ViolentCrime_Harm layer appears in the Contents pane.

14   Adjust your layer visibility so that only the following layers are turned on: ViolentCrime_Harm, ViolentCrime_Optimized, ChicagoPoliceDistricts, and the basemap.

5c

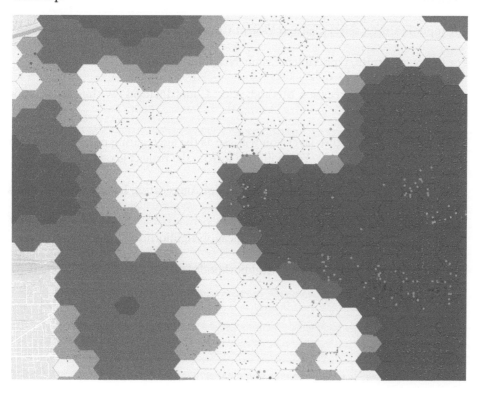

The ViolentCrime_Harm layer has two primary differences from the ViolentCrime_Optimized layer. First, because points do not equally contribute to the calculation, hot and cold spots are more geographically concise. Second, rather than aggregating points into grids, the violent crime points themselves are assigned hot spot values and associated *p* values. You can use the navigation tools on the Map tab to zoom in on the map to get a clearer view of the points.

Optimize hot spots using aggregation polygons

The Optimized Hot Spot Analysis tool also allows users to customize how points are aggregated for the analysis. Rather than using one of the grid types available in the aggregation method list, you can use a polygon feature layer. This may be helpful when officials are interested in crime concentration at pre-existing administrative levels. You will conclude this exercise by creating an optimized hot spot map by aggregating points into Chicago's residential neighborhoods.

> **Note:** Aggregation polygons must contain at least 30 features.

5c

1   Open the Optimized Hot Spot Analysis tool.

2   For Input Features, select Chicago Violent Crime.

3   For Output Features, type **ViolentCrime_Neighborhoods.**

4   For Incident Data Aggregation Method, select Count Incidents Within Aggregation Polygons.

The neighborhoods feature layer you will use as the aggregating polygon is not currently in the Contents pane. You will load it from the project geodatabase.

5   Click the folder to the right of the Polygons For Aggregating Incidents Into Counts field.

6   Browse to Chapter5.gdb, double-click ChicagoNeighborhoods, and run the tool.

The newly created ViolentCrime_Neighborhoods layer appears in the Contents pane.

7 Adjust your layer visibility so that only the VioimentCrime_Neighborhoods, ChicagoPoliceDistricts, and basemap layers are visible.

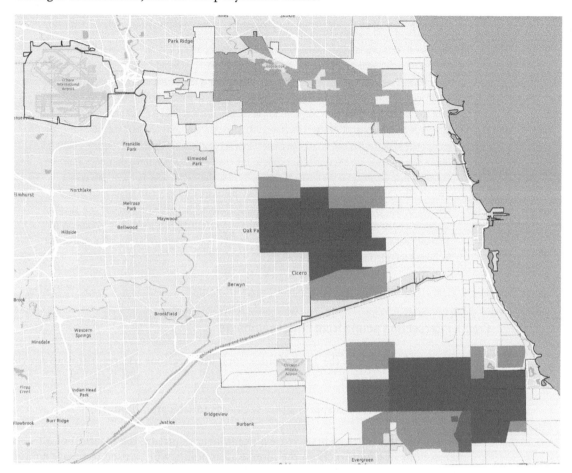

In this map, each neighborhood in Chicago received a z-score and *p* value based on the total number of violent crimes within its boundaries. Calculated this way, more significant hot spots appear than cold spots.

Exercise 5d: Create street intersection points and Thiessen polygons

Research highlight 5d: Using Thiessen polygons in crime-and-place research

Microlevel units of analysis have become increasingly important in crime-and-place research. A focus on small units such as street segments or intersections avoids the ecological fallacy: the incorrect assumption that crime patterns observed at a larger unit apply equally to the microlevel units of which it consists (Johnson et al. 2009). In recent years, crime-and-place scholars have added Thiessen polygons to the catalog of microlevel units of analysis (Haberman 2017; Piza and Gilchrist 2018; Ratcliffe et al. 2011). Thiessen polygons create a Voronoi network using lines to divide a plane into areas closest to a set of points (Chainey and Ratcliffe 2005). Street intersections are often used as points in the creation of such Voronoi networks. Thiessen polygons help account for the inherent ambiguity of spatial crime data. For example, police officers often arbitrarily report crime locations as the nearest intersection rather than a precise address (Braga et al. 2011). Therefore, whether a case is mapped at an intersection or a street segment may depend on officer decision-making rather than the true location of a crime. With Thiessen polygons, crime reported at street intersections is aggregated to the same spatial unit as crimes reported at nearby addresses.

Create intersection points from street segments

1   In the Catalog pane, in the Maps folder, click Intersections to open the map for this exercise.

The Intersections map displays street segments in Chicago, Illinois.

You will use this layer to first create a point layer depicting the location of all street intersections in Chicago. Then the intersection points will be used to create contiguous Thiessen polygons throughout the city.

2   Search for and open the Intersect (Analysis Tools) tool. (**Hint:** On the Analysis tab, click Tools, type **Intersect**, and press Enter.)

The Intersect tool creates a layer reflecting where one or more features overlap. The typical use of the tool is to create a feature class representing where multiple layers intersect. However, for this example, the tool will be used to identify where individual features (street segments) overlap in the Chicago_Streets layer.

3   In the tool pane, for Input Features, select Chicago_Streets.

4   For Output Feature Class, type **Chicago_Streets_Intersect**.

5   For Attributes To Join, select All Attributes.

6 For Output Type, select Point.

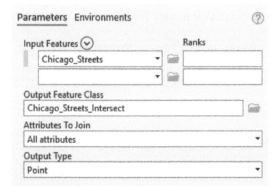

7 Run the tool.

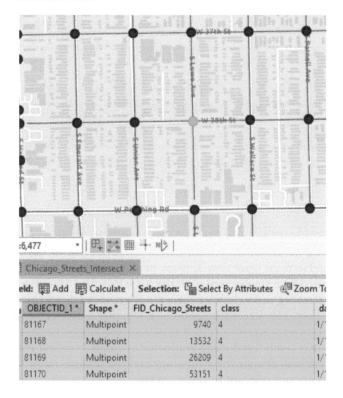

The newly created Chicago_Streets_Intersect layer now appears in the Contents pane. The Intersect tool creates points at each location where a street segment intersects with another. Therefore, individual street intersections are currently counted multiple times. In the figure, four points (one for each of the intersecting street segments) corresponding to the same intersection are selected (shown as a single cyan multipoint).

You will now collapse the data so that each street intersection is represented by a single point.

Collect events to finalize intersection points

1   Search for and open the Collect Events (Spatial Statistics Tools) tool.

2   In the tool pane, for Input Incident Features, select Chicago_Streets_Intersect.

3   For Output Weighted Point Feature Class, type **Street_Intersections**.

4   Run the tool.

5   In the Contents pane, turn off the Chicago_Streets_Intersect layer.

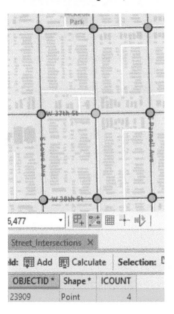

The newly created Street_Intersections layer now appears in the Contents pane. Each street intersection is represented by a single point with the ICOUNT field identifying the number of intersecting street segments. In the figure, the selected point represents a street intersection where four separate street segments come together (ICOUNT = 4).

The Street_Intersections layer is symbolized using graduated symbols to reflect the different ICOUNT values. You will change the symbology style to single symbol because the ICOUNT values are not relevant to this exercise.

6   In the Contents pane, click Street_Intersections.

7   On the Appearance tab, in the Drawing group, click the Symbology button down arrow, and select Single Symbol.

Create Thiessen polygons

You will use the Street_Intersections layer as the point feature in a Voronoi network to create Thiessen polygons.

1 Search for and open the Create Thiessen Polygons (Analysis Tools) tool.

2 In the tool pane, for Input Features, select Street_Intersections.

3 For Output Feature Class, type **Intersection_Polygons**.

4 For Output Fields, select Only Feature ID.

5 Run the tool.

The Intersection_Polygons layer appears in the Contents pane.

5d

6   Turn off the Street_Intersections layer.

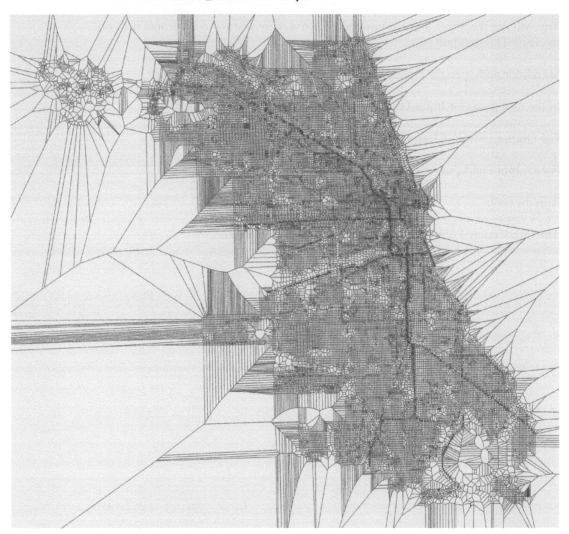

Thiessen polygons are created with no consideration of underlying geography other than the points used to guide their creation. As such, the polygons extend far beyond the study area boundary.

You will use the Clip tool to restrict the Thiessen polygons to the Chicago city boundary.

Clip Thiessen polygons

1   Search for and open the Clip (Analysis Tools) tool.

2   In the tool pane, for Input Features, select Intersection_Polygons.

3 For Clip Features, select Chicago Border.

4 For Output Feature Class, type **Intersection_Polygons_Clip**.

5 Run the tool.

6 In the Contents pane, turn off Intersection_Polygons.

5d

The newly created Intersection_Polygons_Clip layer appears in the Contents pane. Its extent is restricted to the Chicago Border layer.

Exercise 5e: Count incidents and measure percentage changes

Crime analysis highlight 5e: Aggregating points to track crime changes in specialized areas

Joining incidents to areas of interest helps us overcome shortcomings in how our records management system (RMS) stores data. For instance, we generate a weekly COMPSTAT report that breaks down crime trends by bureaus and beats. Our RMS attaches a bureau and a beat to each offense, so counting offenses in these areas is easy. However, I was approached by a city councilmember who wanted us to compute crime trends inside council districts. This is a problem because our offense data does not have a field showing the council district of occurrence. Additionally, council districts cut across bureaus and beats, so simply combining beat or bureau numbers is not an option. Fortunately, we do have city council district GIS layers and adding council district information is easy using the join tools available in ArcGIS Pro. We created a model that takes our offense data out of a geodatabase, geocodes it to a map, performs an incident join, and attaches information from the council district layer to the offense data. The result is a table of offense data that has a column indicating the council district in which each offense occurred. We can also use these tools to track crime changes in council districts over time.

—*Charles Giberti, supervisory crime analyst, Wichita (Kansas) Police Department*

In this exercise, you will measure counts of crime points within polygon features in Chicago. You will use the Summarize Percent Change tool to measure how gun arrest counts changed across districts from the first half of the year to the second half of the year.

Measure percent change

1   In the Catalog pane, in the Maps folder, click Incident Count to open the map for this exercise.

This map displays Chicago police districts and gun arrests occurring during the first (January through June) and second (July through December) halves of the year. Abandoned building and Thiessen polygon layers are also in the map, but their visibility is turned off in the Contents pane.

2   On the Crime Analysis tab, in the Analysis Tools Gallery, click Summarize Percent Change.

3   In the tool pane, for Input Features, select Chicago Police Districts.

4   For Input Current Period Point Features, select Gun Arrests (Jul. – Dec.).

5   For Input Previous Period Point Features, select Gun Arrests (Jan. – Jun.).

6   For Output Feature Class, type **Gun_Arrests_Change**

7   Run the tool.

The newly created Gun_Arrests_Change layer appears in the Contents pane.

8 Turn off both Gun Arrests layers.

9 Right-click the Gun_Arrests_Change layer, and click Label.

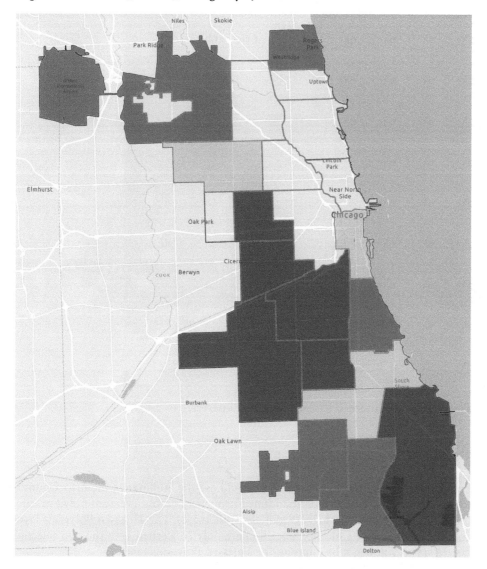

Each district is automatically symbolized by the percentage change in gun arrests between the two periods.

The count and percentage information are included as fields in the attribute table. You can use these fields to learn more about the distribution of your data.

10 Open the attribute table for Gun_Arrests_Change.

11 Right-click the Percent Change field, and click Sort Descending.

The largest change in gun arrests was a 200 percent increase in two separate districts. You will select these districts and zoom to them on the map.

12 Select the first two districts in the table.

13 Click the Zoom To button at the top of the table.

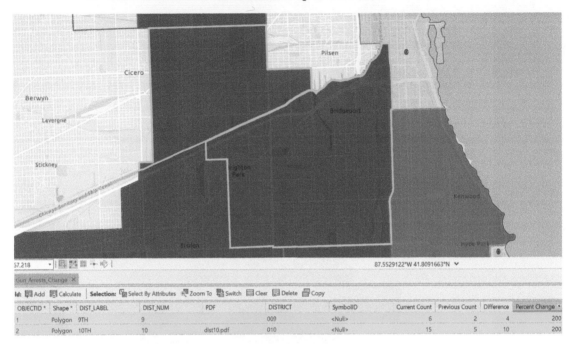

The map is now zoomed to the districts of interest.

Use the Summarize Incident Count tool

For the final portion of this exercise, you will use the Summarize Incident Count tool to measure the distribution of abandoned properties across Thiessen polygons in the two districts with the highest gun arrest increases. The Thiessen Polygons layer covers the entirety of Chicago. Since you are only concerned with the two selected districts, you will begin by clipping the Thiessen Polygons layer.

> **Note:** The next steps will not generate the intended outcome if the two districts are not selected. Make sure that the two districts are selected before proceeding.

1 Open the Clip tool.

2 In the tool pane, for Input Features, select Thiessen Polygons.

3 For Clip Features, select Gun_Arrests_Change.

4 For Output Feature Class, type **Thiessen_Polygons_Clip**.

5 Run the tool.

The Thiessen_Polygons_Clip layer appears in the Contents pane.

6 Turn off the Gun_Arrests_Change layer.

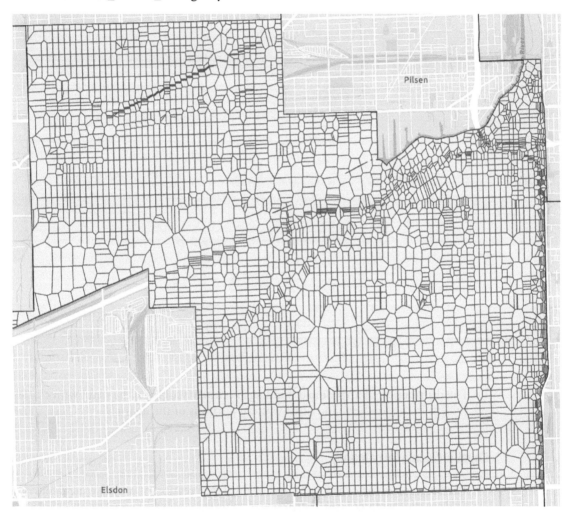

7 Search for and open the Summarize Incident Count tool.

8 In the tool pane, for Input Features, select Thiessen_Polygons_Clip.

9 For Input Summary Features, select Chicago Abandoned Buildings.

10 For Output Feature Class, type **Thiessen_Abandoned**.

11 Run the tool.

The distribution of abandoned buildings is clearly visible on the map, with clusters evident in the southern and northwestern portions of the districts.

Count incidents by category

The Summarize Incident Count tool enables the creation of multiple incident count fields in the resulting layer's attributes. In this step, you will conduct an incident count based on the status (boarded or open) of the abandoned building.

1 Reopen the Summarize Incident Count tool.

2 For Input Features, select Thiessen_Polygons_Clip.

3 For Input Summary Features, select Chicago Abandoned Buildings.

4 For Output Feature Class, type **Thiessen_Abandoned_Type**.

5 For Group Field, select IS_BUILDIN.

6 Run the tool.

The Thiessen_Abandoned_Type layer appears in the Contents pane. Visually, the layer looks identical to the prior incident count output. However, the attribute table contains more information on the abandoned buildings. The Total Count field lists the total number of abandoned buildings within the polygon. Three additional fields provide counts of the types of abandoned buildings in the polygon: Open, Boarded, and an unnamed field denoting the buildings for which the IS_BUILDIN field is blank.

OBJECTID *	Shape *	Input_FID		Boarded	Open	Total Count ▾	Shape_Length	Shape_Area
1896	Polygon ZM	20040	0	0	6	6	0.006054	0.000002
209	Polygon ZM	8249	0	0	5	5	0.006022	0.000002
1833	Polygon ZM	18479	1	0	4	5	0.006053	0.000002
1851	Polygon ZM	19025	0	0	5	5	0.00545	0.000002
2060	Polygon ZM	22544	0	1	4	5	0.006278	0.000002

Summary

This chapter taught you how to use various tools to identify geographic hot spots, and to measure changes in the data. The latter task is often ignored by police agencies, including command staff that should dutifully measure change in activity. Some of the newest ArcGIS Pro tools provide cutting-edge data analysis needed for informed decision-making in modern policing. Keep in mind that hot spots of crime are not the only hot spots worthy of attention. Emergency (911) calls for service, officers' self-initiated activity, or even citizen complaints can all be measured using the tools covered in this chapter for increased awareness.

5e

Chapter 6
Incorporating time in spatial analysis

Overview

In this chapter, you will learn how to use the spatiotemporal analysis tools in ArcGIS Pro. Crime analysts have been adept at readily identifying the *where* and *what* through GIS application for several years. An often-overlooked aspect of crime analysis is the *when*. By simultaneously accounting for space and time, these techniques can help refine crime analysis outputs and provide actionable intelligence to police commanders who need to know *when* to deploy crime prevention resources as well as *where* such resources should be deployed.

You will acquire the following skills upon completing the exercises in this chapter:
- Creating and visualizing space-time cubes
- Measuring crime trends over time
- Creating sequence lines and points
- Labeling sequence points
- Creating incident paths
- Creating and exporting animations

Download and install the data

Before working on the exercises, you will obtain the necessary data.

1   Go to www.arcgis.com and sign in with your ArcGIS Online account credentials.

2   Type **Modern Policing Using ArcGIS Pro (Esri Press)** in the search box, and click the Groups tab. Make sure that the Only Search In Your Organization option is turned off.

3   Click the Modern Policing Using ArcGIS Pro (Esri Press) group.

4   On the group heading, click Content.

5   Click the chapter 6 file and download it.

6   Locate the zip file you downloaded to your local drive, right-click, and extract it to C:\ModernPolicing.

This will create a folder named Chapter6 in the ModernPolicing folder.

This folder contains an ArcGIS Pro project and data you will use for the exercises in this chapter.

Exercise 6a: Create space-time cubes and identify emerging hot spots

A space-time cube aggregates points into space-time bins to evaluate how counts and summary field values change over time. Incident data is analyzed within each bin of a cube and any temporal-based activity or trend is measured across time.

It is important to review key requirements of the Create Space Time Cube By Aggregating Points tool before you begin this portion of the exercise. First, it is recommended, but not required, that your data be divided into equal time intervals, such as weeks, months, or years depending on the amount of data and nature of the analysis you will perform. The tool can leave any remaining data in a separate bin in the space-time cube, but this may yield incomplete results. For example, using a one-week time interval with 90 days of data would result in 12 full weeks with six days remaining in the final data bin. Second, a space-time cube must have at least 10 time slices. You could not, for example, analyze 60 days of data using a one-week time interval. Instead, you would need to use a smaller time interval, such as four days. Finally, the tool requires data to be in a projected coordinate system.

Create a space-time cube

1   Open Chapter6.aprx in C:\ModernPolicing\Chapter6.

The project shows a map named Space-Time. The map displays locations of motor vehicle theft incidents in Rochester, New York, from 2017 to 2019. You will use this layer to create a space-time cube and identify emerging hot spots.

2   On the Crime Analysis tab, in the Analysis Tools gallery, click Create Space Time Cube.

3   In the tool pane, for Input Features, select Motor Vehicle Theft (2017_2019).

4   For Output Space Time Cube, in C:\ModernPolicing\Chapter6, type **MVT_Cube**.

5   For Time Field, select Occurred_3.

6   For Time Step Interval, type **1**, and select Months for the interval.

7   For Time Step Alignment, select End Time.

8   For Aggregation Shape Type, select Hexagon Grid.

9   For Distance Interval, type **1000**, and select Feet for the interval.

10 Run the tool.

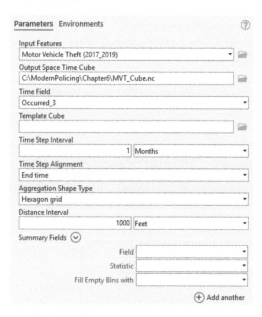

> **Note:** You can use a pre-existing polygon feature to establish the boundary for the point aggregations. For example, you can aggregate points to police sectors or precincts rather than a contiguous grid. This may be useful for identifying police commands that need additional crime prevention resources and other purposes.

As opposed to most other geoprocessing tools in ArcGIS Pro, the Create Space Time Cube By Aggregating Points tool does not add new layers to the Contents pane. Rather, output from this tool is a netCDF representation of your input points, which can be used as input for other tools such as the Emerging Hot Spot Analysis tool. You can visualize the space-time cube in the Contents pane by using the Visualize Space Time Cube tool.

Visualize the space-time cube

1 Search for and open the Visualize Space Time Cube In 2D tool.

2 In the tool pane, for Input Space Time Cube, click the folder to the right of the text box, browse to C:\ModernPolicing\Chapter6, and double-click MVT_Cube.nc.

3 For Cube Variable, select COUNT.

4 For Display Theme, select Trends.

5 For Output Features, type **MVT_Cube_trends**.

6a

6 Check the Enable Time Series Pop-ups check box.

7 Run the tool.

8 Turn off the Motor Vehicle Theft (2017_2019) layer.

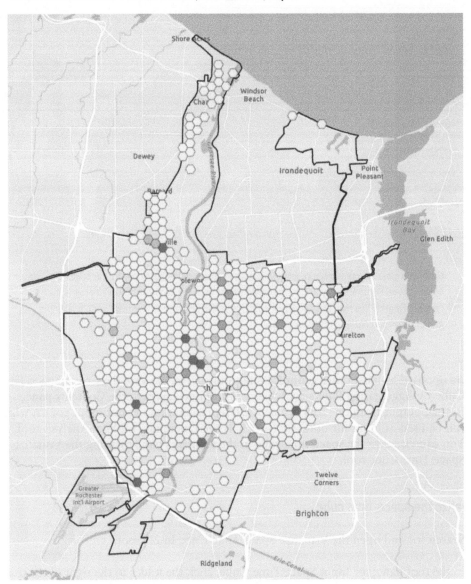

The new cube trend layer appears in the Contents pane. The layer is symbolized according to the motor vehicle theft trend observed within the grids over the 36-month period. This layer identifies layers for which trends were identified, whether the trend was upward (crime increase) or downward (crime decrease), and the statistical significance of the trend.

The attribute table of this layer contains the z-score and associated *p* values. Clicking the northeastern-most green grid with the Explore tool displays its attributes, including the z-score and *p* value.

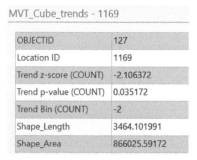

MVT_Cube_trends - 1169

OBJECTID	127
Location ID	1169
Trend z-score (COUNT)	-2.106372
Trend p-value (COUNT)	0.035172
Trend Bin (COUNT)	-2
Shape_Length	3464.101991
Shape_Area	866025.59172

Create an emerging hot spot map

With the space-time cube created, you will now create an emerging hot spot. This will yield a practical visualization of spatiotemporal trends in the motor vehicle theft data.

1   On the Crime Analysis tab, in the Analysis Tools gallery, click Emerging Hot Spot.

2   In the Emerging Hot Spot Analysis tool pane, for Input Space Time Cube, click the folder to the right of the text box.

3   Browse to C:\ModernPolicing\Chapter6, and double-click MVT_Cube.nc.

4   For Analysis Variable, select COUNT.

5   For Output Features, type **MVT_Emerge**.

6   For Conceptualization Of Spatial Relationships, select Fixed Distance.

> **Note:** The Conceptualization Of Spatial Relationships parameter should be selected in consideration of the inherent relationship between the features being analyzed. For best practices in conceptualizing spatial relationships, see ArcGIS Pro > Tool Reference > Geoprocessing Tools > Spatial Statistics Toolbox > Modeling Spatial Relationships > Selecting A Conceptualization Of Spatial Relationships: Best Practices.

7   For Neighborhood Distance, type **1000**, and select Feet for the measurement.

8   For Number Of Spatial Neighbors, type **6**.

> **Note:** Six spatial neighbors were specified to correspond to the number of sides in a hexagon grid.

9 For Neighborhood Time Step, type **1**.

10 For Polygon Analysis Mask, select Rochester_Border.

11 Run the tool.

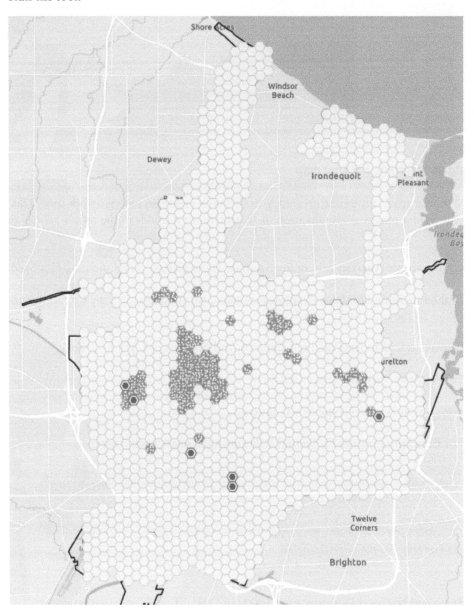

The new MVT_Emerge layer appears in the Contents pane. The layer is symbolized with 17 unique values corresponding to the hot spot status of the individual grids. A number of grids are identified as sporadic hot spots while six are identified as new hot spots. The remainder of grids are categorized as no pattern detected.

6a

Exercise 6b: Map incident sequences

Crime analysis highlight 6b: Analyzing a crime series

There are two primary ways an analyst will identify a crime series—through either inductive or deductive means. By scouring a large database for specific patterns, an analyst is engaging in deductive reasoning. Alternatively, if an analyst is reading through several reports and notices a pattern developing, they are engaging in inductive reasoning. Inductively, one of the most typical ways to identify a crime series is reading through various reports, identifying commonalities, and then mapping the sequence of the pattern using points, documenting the specific location of the incident along with any other relevant detail that would not clutter the map. From there, an analyst would attempt to determine any further pattern similarities, including conducting a time analysis for the pattern and potentially forecasting when the next incident in the pattern would occur. Alternatively, an analyst could look at a larger set of points on a map and apply a series of spatial queries or filters that would help identify spatial clusters or some type of spatial pattern. Such maps would be included as part of a tactical crime analysis bulletin that would document the who, what, where, when, and how of crime.

— *Special Constable John Ng, divisional crime analyst, Saskatoon, Saskatchewan (Canada) Police Service*

Create sequence lines and points

1   In the Catalog pane, in the Maps folder, click Sequence to open the map for this exercise.

The map displays carjacking incidents in Santa Rosa, California. A layer containing burglary of automobile events also appears in the Contents pane with its visibility turned off. You will use these feature layers to work with the Points To Track Segments tool, which creates connecting lines and outputs showing the sequential order of related incidents. This information can clarify the nature of an observed crime series.

2   Search for and open the Points To Track Segments (Intelligence Tools) tool.

3   In the tool pane, for Input Features, select Carjackings Santa Rosa.

4   For Date Field, select Date.

5   For Output Feature Class, type **Carjackings_Sequence**.

6   Uncheck the Include Velocity Fields check box.

6b

7   For Output Sequence Points, type **Carjackings_Points.**

8   Click Run.

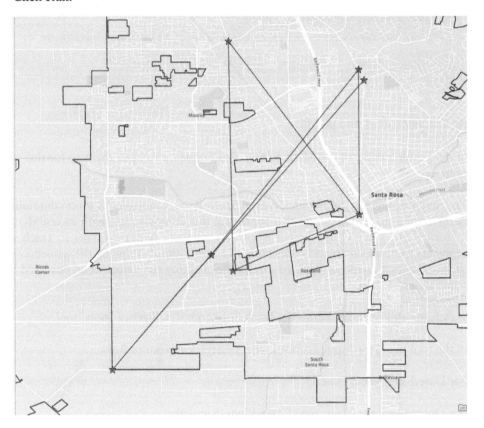

The Carjackings_Sequence and Carjacking_Points layers appear in the Contents pane.

The points in the Carjackings_Points layer correspond to features in the Carjackings Santa Rosa layer used in the Points To Track Segments tool. The attribute table for

Carjackings_Points includes a field named Sequence, denoting the sequential order of each feature.

OBJECTID *	Shape *	sequence	Date
1	Point	4	1/4/2019
2	Point	3	2/25/2019
3	Point	8	4/18/2019
4	Point	5	8/23/2019
5	Point	1	10/29/2019
6	Point	7	11/2/2019
7	Point	6	11/9/2019
8	Point	2	11/15/2019

Carjackings_Sequence is a line feature reflecting the sequential ordering of all eight carjacking incidents. This layer creates a path among all data points in the Carjackings Santa Rosa layer. The attribute table for Carjackings_Sequence notes the distance between the sequential points in the Carjackings Santa Rosa layer.

> **Note:** The Points To Track Segments tool assumes that an individual (or group of individuals) traveled directly from point to point during a single trip. Consequently, the attribute table includes fields denoting the end date of the trip (the date of the next incident) and the temporal duration (measured in minutes and seconds) between the start and end dates. These fields can be ignored when individuals are not assumed to have traveled in a single trip.

Create sequence lines and points by group

6b

The prior example treated all points in the feature layer as part of the same crime series. Often, you work with data consisting of incidents from a range of different series. In such cases, you can use an option in the Points To Track Segments tool to create multiple line and point features corresponding to the different series contained within your data.

1   Turn off the visibility for Carjackings Santa Rosa, Carjackings_Sequence, and Carjackings_Points.

2   Turn on the visibility for Burglary From Auto.

Burglary From Auto displays the number of burglary from auto incidents from 7/1/2019 to 7/19/2019. You will run the Points To Track Segments tool on this data but will treat incidents within the same police zone as their own series. This will result in the creation of multiple sequence lines to reflect the different series contained in the single feature layer.

3   Reopen the Points To Track Segments tool.

4   For Input Features, select Burglary From Auto.

5   For Date Field, select Date.

6   For Output Feature Class, type **Burglary_Auto_Group**.

7   For Group Field, select Zone.

8   For Output Sequence Points, type **Burglary_Points**.

9   Uncheck the Include Velocity Fields check box.

10 Click Run.

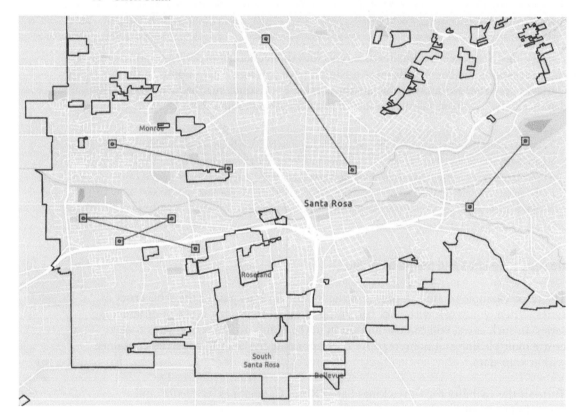

Like the prior example, the Points To Track Segments tool creates two outputs: a line feature (Burglary_Auto_Group) and a point feature (Burglary_Points).

The attribute table is similar to the prior example, with some small differences. Burglary_Auto_Group contains a field for the Zone that each track segment falls within.

In Burglary_Points, a sequence field is added, but the sequence field values reflect the incident order in their specific Zone rather than the layer as a whole.

Exercise 6c: Create incident paths

Crime analysis highlight 6c: Using incident paths for criminal investigations

Property crimes are often difficult to solve. However, when a path can be created from the event to where stolen items end up, a new analysis can occur. Take motor vehicle thefts, for instance. Mapping where a vehicle is stolen and recovered can tell a story. Linking the events, taking the amount of time into consideration, reveals a lot about the offender and the motive (whether the theft was for transportation, a joyride, or a new car). Being able to link surrounding events can contribute to the analysis. When the vehicle was recovered, was there another vehicle stolen in the area around the same time? The same methods can be applied when connecting an auto burglary series. Auto burglars tend to discard unwanted property along the way on a spree. Linking the property from event to event generates a path taken by the criminal on the specific day. For example, a burglar broke into a car in a city in Northern California. Property from that burglary was discarded at another auto burglary site along the US 101 corridor. Finally, property from multiple crimes scenes was located in a dumpster at the final burglary of the day. The routes the offender takes are not always known to the analyst, but when creating a connection between two or more incidents, an event path is created. That can begin paving the way to solving the generally unsolvable.

— *Jillian Goshin, crime and intelligence analyst, San Rafael (California) Police Department*

Create incident paths between points

1 In the Catalog pane, in the Maps folder, click Incident Path to open the map for this exercise.

The map displays locations of vehicle thefts and their corresponding recovery locations in the eastern portion of São Paulo, Brazil. A layer displaying vehicle theft hot spots grids, identified by the Optimized Hot Spot tool, is also in the Contents pane with its visibility turned off.

You will use these layers to work with the Generate Origin-Destination Links tool, which creates connecting lines between the locations where crime events originate and conclude—in this example, where motor vehicles were stolen and ultimately recovered. This information can clarify an offender's direction of flight after committing a crime.

2 Search for and open the Generate Origin-Destination Links (Analysis Tools) tool.

3 In the tool pane, for Origin Features, select Vehicle Theft.

4 For Destination Features, select Vehicle Recovery.

5 For Output Feature Class, type **Theft_Recovery_Path**.

6 For both Origin Group Field and Destination Group Field, select TARGET_FID.

7   For Line Type, select Planar.

8   Run the tool.

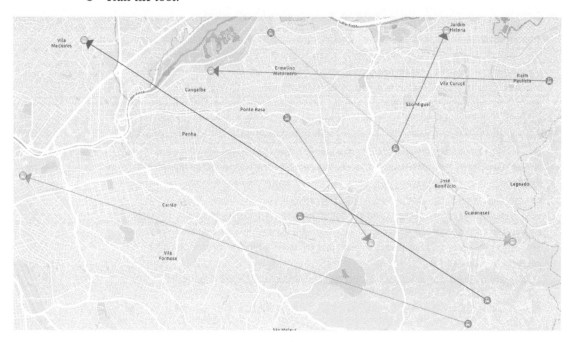

The Theft_Recovery_Path layer appears in the Contents pane. It includes seven individual features, one for each of the theft and recovery links in the data.

The Generate Origin-Destination Links tool also creates charts displaying the sum and mean distances between the origin and destination points.

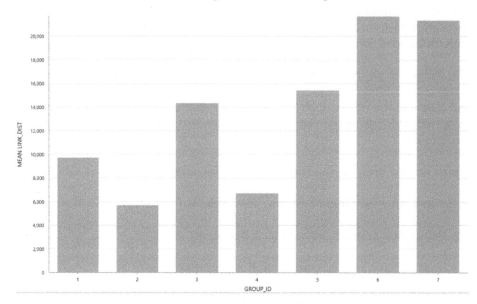

The value shown in the LINK_DIST field is the same as the unit of measurement used in the current projection. Distance between points is reflected in meters, given the use of the WGS coordinate system in the current project. It may be more informative to measures distance paths in another distance unit, depending on the audience.

Calculate distance between origin and destination points

1   Open the attribute table for Theft_Recovery_Path.

    A variety of information is included in the attribute table, such as coordinates of the origin and destination for each theft. You will add a new field that measures the miles between points.

2   At the top of the attribute table, click the Add Field button.

3   In the Fields table, for both the Field Name and the Alias fields, type **Miles**.

4   For Data Type, select Double.

5   Save your edits and close the Fields table.

6   In the attribute table, right-click the Miles field, and select Calculate Geometry.

7   In the Calculate Geometry pane, ensure that Input Features is set to Theft_Recovery_Path.

8   Under Geometry Property, for Property, select Length (Geodesic).

9   For Length Unit, select Miles (United States).

10 Run the tool.

Miles
5.524281
3.224438
8.140279
3.805822
8.776052
12.333893
12.145793

The attribute table now includes a Miles field measuring the distance between the origin (vehicle theft) and destination (vehicle recovery) points.

Note: You can also determine the distance unit in the Geoprocessing pane when running the tool.

6c

Create incident paths between polygons and points

In addition to creating paths between separate point features, the Generate Origin-Destination Links tool allows users to create paths between polygons and points. You will conclude this exercise by creating connecting lines between preidentified vehicle theft hot spot grids and the vehicle recovery points.

1   In the Contents pane, turn off the Vehicle Theft and the Theft_Recovery_Path layers.

2   Turn on the Vehicle Theft Grids layer.

This layer displays four features of merged hexagon grids created with the Optimized Hot Spots tool. Three features (named *H.S.*) consist of contiguous hexagon grids identified as significant crime hot spots. One feature (named *N.S.*) consists of contiguous hexagon grids that did not achieve statistical significance (in other words, they are neither a hot spot nor a cold spot).

You will create connecting lines between the grids where vehicle thefts occurred and vehicle recovery points.

3   Open the Generate Origin-Destination Links tool.

4   For Origin Features, select Vehicle Theft Grids.

5   For Destination Features, select Vehicle Recovery.

6   For Output Feature Class, type **Grid_Path**.

7   For both Origin Group Field and Destination Group Field, select Grid.

8   For Line Type, ensure that Planar is selected.

6c

9 Run the tool.

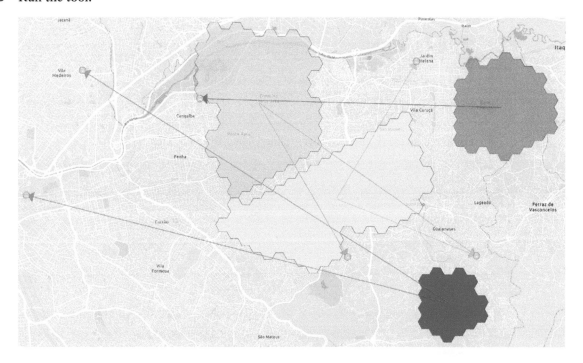

The Grid_Path layer and associated charts appear in the Contents pane. The connections between the grid centroids and the recovery points allow for the visualization of specific patterns. For example, the two vehicles stolen within H.S. #1 (yellow) traveled in different directions. Furthermore, the two southeasternmost recovered vehicles were stolen from two different grids (H.S. #1 and N.S. #1).

Exercise 6d: Use the time slider to create time animations

Animations can help you tell a story by showing how a map feature changes over time. Animations can be modified interactively, supplemented with overlay text and images, and exported as a video file for sharing.

Set time properties

1 In the Catalog pane, in the Maps folder, click Animation to open the map for this exercise.

This map displays locations where extra police patrol units were stationed in Santa Rosa, California, from 3/1/20 to 6/30/20. You will use this data to animate the locations of extra patrols over the six-month period.

Before animating a layer, you must set its time properties.

2 Open the layer properties window for Extra Patrol Santa Rosa. (**Hint:** In the Contents pane, double-click Extra Patrol Santa Rosa.)

3 Click the Time tab.

4 For Layer Time, select Each Feature Has A Single Time Field.

5 For Time Field, select Date.

6 For Time Interval, click No Pre-defined Time Interval, and click OK.

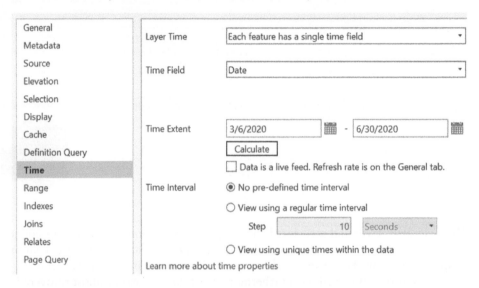

A contextual Time tab appears in the ribbon now that the extra patrol layer is designated as time data.

Create an animation

1 On the View tab, in the Animation group, click Add.

Using the Animation tool, you will create keyframes to customize the display shown across different portions of the animation. For basic animations, only two keyframes are needed: one for the beginning of the time period and one for the end of the time period.

2   In the map view, move the time slider to the beginning of the time period.

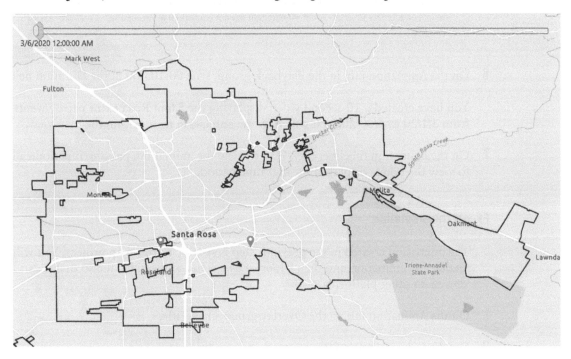

You will add this data view as the first keyframe in the animation.

3   On the Animation Tab, in the Create group, click Append.

You can also add keyframes by using the tools directly within the Animation Timeline pane.

4   In the map view, move the time slider to the end of the time period.

5   In the Animation Timeline pane, click the green plus (+) button to append the next keyframe.

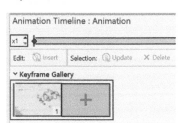

> **Note:** If you do not see the Animation Timeline pane, make sure that you activated it on the Animation tab. The Time tab and the Animation tab are contextual in ArcGIS Pro.

6 On the Animation tab, in the Playback group, type **00:10.000** in the Duration field.

You have created a 10-second video displaying the Santa Rosa extra patrol events from 3/1/20 to 6/30/20. You will view the animation from within the map view.

7 On the animation timeline, click Play to view the animation, pausing it periodically to view the points displayed for that time period.

Export the animation

The animation created in the previous step is specific to this map project. You will finalize the animation and then export it as a stand-alone video file so that it can be viewed on other platforms.

1 On the Animation tab, in the Overlay group, click Title.

2 For the font size, select 20.

3 In the Title text box, type **Extra Patrol: March – June**.

4 Click the red *X* to close the Title window.

The title appears at the top of the animation.

5 On the Animation tab, in the Export group, click the Export Movie button.

The Export Movie pane provides several presets designed to maximize the animation parameters for the intended platform. For this exercise, you will leave the settings as is.

6 In the Export Movie pane, click the folder to the right of the File Name field.

7 Browse to C:\ModernPolicing\Chapter6\Animations.

8 For File Name, type **ExtraPatrol_Animation**.

9 Click Save.

10 Click Export.

ExtraPatrol_Animation is now a stand-alone video file on your computer.

6d

Summary

This chapter taught you skills related to an aspect of GIS applications (including crime analysis) that is too often overlooked: time. The influence of time is often overshadowed because of the inherent spatial nature of GIS data and applications. ArcGIS Pro helps overcome that limitation with tools to create space-time cubes, create and label the sequence of related events, and animate data, all of which were discussed in this chapter. Bringing the dimension of *when* into your analysis will enhance the influence of your work by maximizing its practical value for police commanders and other decision-makers.

Chapter 7
Using spatial statistics to identify spatial relationships

Overview

In this chapter, you will learn how to use a range of spatial statistics tools in ArcGIS Pro. These tools depart from more traditional statistical methods in the sense that they were developed specifically for use with geographic data. By incorporating space directly into their mathematical algorithms, these tools provide optimal performance when analyzing spatial patterns and relationships between variables.

You will acquire the following skills upon completing the exercises in this chapter:
- Measuring the colocation of point features
- Identifying density-based clusters
- Adjusting algorithms to customize selection parameters
- Identifying multivariate attribute clusters
- Performing generalized linear regression
- Performing geographically weighted regression
- Interpreting regression outputs and model fit statistics

Download and install the data

Before working on the exercises, you will obtain the necessary data.

1   Go to www.arcgis.com and sign in with your ArcGIS Online account credentials.

2   Type **Modern Policing Using ArcGIS Pro (Esri Press)** in the search box, and click the Groups tab. Make sure that the Only Search In Your Organization option is turned off.

3   Click the Modern Policing Using ArcGIS Pro (Esri Press) group.

4   On the group heading, click Content.

5   Click the chapter 7 file and download it.

6 Locate the zip file you downloaded to your local drive, right-click, and extract it to C:\ModernPolicing.

This will create a folder named Chapter7 in the ModernPolicing folder.

This folder contains an ArcGIS Pro project and data you will use for the exercises in this chapter.

Exercise 7a: Measure colocation of points

The Colocation Analysis tool applies a statistical algorithm to determine the level to which points are colocated or isolated in space. For each feature, the tool calculates a local colocation quotient (LCLQ) to reflect the level to which it colocates with neighboring features. An LCLQ value greater than 1 means that the feature is more likely to be near another feature; an LCLQ less than 1 means the feature is less likely to be near another feature. The tool provides several options for measuring the spatial relationship between features.

Measure colocation with K nearest neighbors

1 Open Chapter7.aprx in C:\ModernPolicing\Chapter7.

The project shows a map named Colocation. The map displays locations of auto theft and theft from auto incidents in 2019 in Indianapolis, Indiana. A visual inspection of the points also highlights certain areas with high levels of both crime types.

You will use this data to perform a colocation analysis to measure the spatial association between the auto theft points and the theft from auto points.

2 On the Analysis tab, click Tools.

3 In the Geoprocessing pane, search for and open the Colocation Analysis tool.

4 In the Colocation Analysis tool pane, for Input Type, select Two Datasets.

5 For Input Features Of Interest, select IMPD Auto Theft 2019.

6 For Input Neighboring Features, select IMPD Theft From Auto 2019.

> **Note:** It is possible to use colocation analysis on a single layer if your data contains attribute fields specifying the features of interest and neighboring features.

7 For Neighborhood Type, select K Nearest Neighbors.

8   For Number Of Neighbors, type **10**.

9   For Number Of Permutations, select 999.

> **Note:** Permutations are used to calculate p-values for the LCLQ measures. The higher the number of permutations, the more precise the measure of statistical significance. However, selecting a higher number of permutations increases the processing time.

10   For Output Features, type **Auto_Colocation**.

Input Type
Two datasets ▾

Input Features of Interest
IMPD Auto Theft 2019 ▾ 📁

Field of Interest
▾

Time Field of Interest
▾

Category of Interest

Input Neighboring Features
IMPD Theft From Auto 2019 ▾ 📁

Field Containing Neighboring Category
▾

Time Field of Neighboring Features
▾

Neighboring Category

Neighborhood Type
K nearest neighbors ▾

Number of Neighbors
10

Number of Permutations	999 ▾

Output Features
Auto_Colocation 📁

> **Additional Options**

7a

11   Run the tool.

The new Auto_Colocation layer appears in the Contents pane.

12   Turn off the IMPD Auto Theft 2019 and IMPD Theft from Auto 2019 layers.

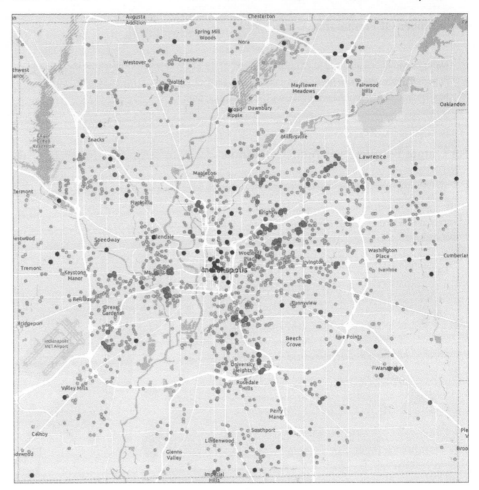

This layer adds the results from the colocation analysis to the IMPD Auto Theft 2019 points. Points are symbolized by their spatial relationship to neighboring features (colocated, isolated, or undefined) as well as the statistical significance of the local colocation quotient (LCLQ).

The Colocation Analysis tool also creates a chart that graphically displays the relationship between the LCLQ and the associated *p*-value. Points in the chart use the same symbology as the feature layer, allowing for the visual identification of each LCLQ category.

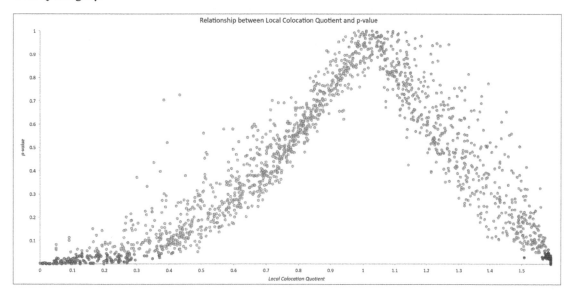

The attribute table for the Auto_Colocation layer contains fields for Local Colocation Quotient, p-value, LCLQ Bin, and LCLQ Type for each of the auto theft features.

OBJECTID *	Shape *	Source ID	Local Colocation Quotient	p-value	LCLQ Bin	LCLQ Type
1	Point	1	0.832349	0.628	3	Isolated - Not Significant
2	Point	2	1.504128	0.126	1	Colocated - Not Significant
3	Point	3	0.386903	0.174	3	Isolated - Not Significant
4	Point	4	1.030971	0.98	1	Colocated - Not Significant
5	Point	5	1.18302	0.64	1	Colocated - Not Significant

The tool allows users to select various methods for measuring spatial relationships as well as add a temporal component to the LCLQ measurement. You will explore these methods over the remainder of this exercise.

Measure colocation with a distance band

1 Open the Colocation Analysis tool.

2 For Input Type, select Two Datasets.

3 For Input Features Of Interest, select IMPD Auto Theft 2019.

4 For Input Neighboring Features, select IMPD Theft From Auto 2019.

7a

5 For Neighborhood Type, select Distance Band.

6 For Distance Band, type **2000**, and select Feet as the measurement unit.

7 For Number Of Permutations, select 999.

8 For Output Features, type **Auto_Colocation_Distance**.

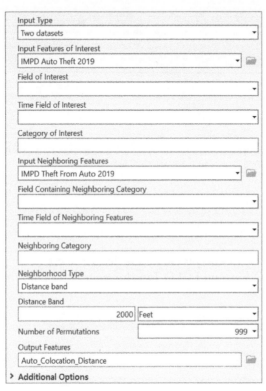

9 Run the tool.

The new Auto_Colocation_Distance layer appears in the Contents pane.

10 Turn off the Auto_Colocation layer.

Compared with the K-means method, using a distance of 2,000 feet in the LCLQ calculation results in the significant colocated points being more highly concentrated in the center portion of the city. Clusters of significant isolated points are also more readily evident to the immediate east and southeast of the central colocated cluster.

Measure colocation with a temporal relationship

You will conclude this exercise by adding a temporal parameter to the colocation analysis.

1 Open the Colocation Analysis tool.

2 In the Colocation Analysis tool pane, for Input Type, select Two Datasets.

3 For Input Features Of Interest, select IMPD Auto Theft 2019.

4 For Time Field Of Interest, select DATE_.

7a

5   For Input Neighboring Features, select IMPD Theft From Auto 2019.

6   For Time Field Of Neighboring Features, select DATE_.

7   For Neighborhood Type, select Distance Band.

8   For Distance Band, type **2000**, and select Feet as the measurement unit.

9   For Temporal Relationship Type, select Span.

> **Note:** Users can also specify the temporal relationship as either "before" or "after." With these settings, the time step extends back in time or forward in time for each input feature of interest. With the span option, the time window extends both back and forward in time for each input feature of interest.

10 For Time Step Interval, type **30**, and select Days as the measurement unit.

11 For Number Of Permutations, select 999.

12 For Output Features, type **Auto_Colocation_Time**.

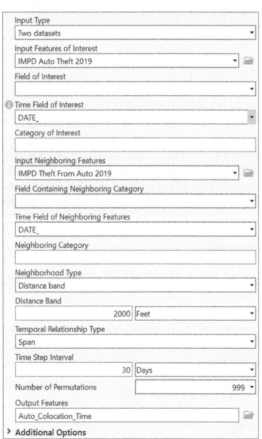

13 Run the tool.

The new Auto_Colocation_Time layer appears in the Contents pane.

14 Turn off the Auto_Colocation_Distance layer.

Introducing the 30-day time span into the analysis results in significantly colocated incidents being even more concentrated in the central portion of the city.

Exercise 7b: Analyze patterns with density-based clustering

In this exercise, you will use density-based clustering tools to identify unique clusters of points throughout the Shotspotter coverage area in Washington, DC. Other density and hot spot identification tools identify aggregate geographies where points concentrate, but density-based clustering applies machine learning algorithms to identify spatial concentrations and identify typologies of clusters in your data. Individual points are assigned to specific clusters based on their spatial relationship to all other point features in the feature class.

Density-based clustering can be run with one of three machine learning processes: DBSCAN, HDBSCAN, or OPTICS.

7b

Identify clusters with DBSCAN

1   In the Catalog pane, in the Maps folder, click DB Clusters to open the map for this exercise.

    This map displays gunshots detected by the Washington, DC, Shotspotter system in 2018.

2   Search for and open the Density-based Clustering (Spatial Statistics) tool.

3   In the tool pane, for Input Point Features, select Shotspotter Gunshots 2018.

4   For Output Features, type **Gunshots_DB**.

5   For Clustering Method, select Defined Distance (DBSCAN).

6   For Minimum Features Per Cluster, type **25**.

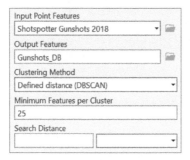

    For this step, you will leave the search distance blank to allow the DBSCAN algorithm to specify a search distance for the cluster identification.

7   Run the tool.

    The new Gunshots_DB layer appears in the Contents pane.

8   Ensure that only these layers are visible: Gunshots_DB, DC Boundary, and the basemap.

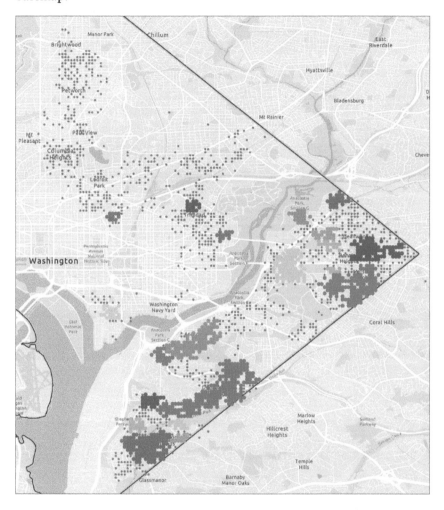

In total, 20 unique clusters of gunshot incidents were identified by the DBSCAN process. Points that do not belong to any clusters are classified as Noise.

> **Note:** Each color used in the symbology is assigned to multiple clusters. Colors are assigned and repeated in a manner that ensures that each cluster is visually distinct from its neighboring clusters.

A chart displaying the features per cluster also appears in the Contents pane. The chart is a bar graph showing each cluster ID and the number of points it contains. Cluster ID 1 pertains to the points classified as Noise (not belonging to any identified cluster). The chart is interactive with the map. Note that cluster ID 4 is the largest. You will see where on the map this cluster's points fall.

7b

9 Open the chart, and click Cluster ID 4.

The corresponding points are selected on the map.

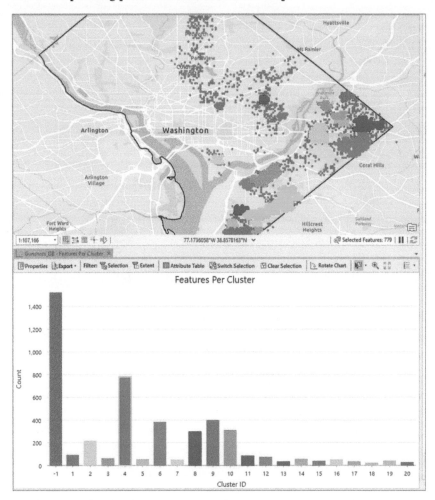

The attribute table for Gunshots_DB contains a field identifying the cluster to which the point belongs (Cluster ID) and the color used in the symbology.

10 Open the attribute table and view the records.

OBJECTID *	Shape *	Source ID	Cluster ID	Color ID
1	Point	1	-1	-1
2	Point	2	1	2
3	Point	3	4	8
4	Point	4	2	3
5	Point	5	9	2

11 Clear all selected records.

Use DBSCAN with a specified search distance

In the prior example, the DBSCAN algorithm automatically specified a search distance for identifying and classifying clusters. However, many times you will have a search distance in mind that best reflects the purpose of your analysis. In such cases, you can specify the search distance for DBSCAN to use in identifying clusters.

1 Open the Density-based Clustering tool.

2 For Input Point Features, select Shotspotter Gunshots 2018.

3 For Output Features, type **Gunshots_DB_distance**.

4 For Clustering Method, select Defined Distance (DBSCAN).

5 For Minimum Features Per Cluster, type **25**.

6 For Search Distance, type **750** and select Feet as the measurement unit.

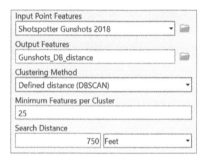

7 Run the tool.

The new Gunshots_DB_distance layer appears in the Contents pane.

7b

8 Turn off the Gunshots_DB layer.

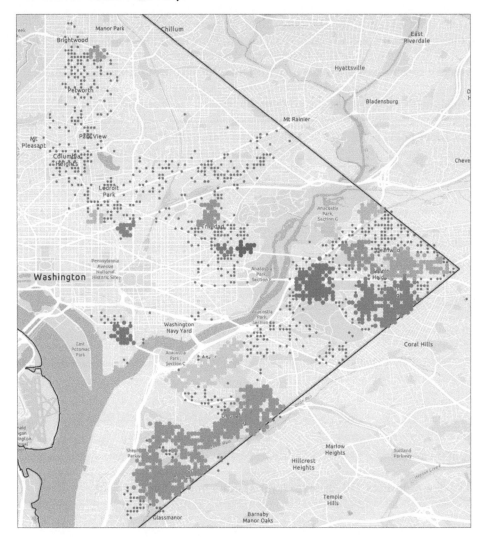

In total, 15 unique clusters of gunshot incidents were identified. There are some key differences between this layer and Gunshots_DB, which was created without a specified search distance. For one, many of the individual clusters identified in Gunshots_DB were aggregated into single, larger clusters in Gunshots_DB_distance. More clusters were identified in the northern area of Washington, DC, in Gunshots_ DB_distance than in Gunshots_DB.

Identify clusters with HDBSCAN

HDBSCAN differs from DBSCAN in the sense that it is a hierarchical, self-adjusting cluster identification method. HDBSCAN requires users to specify only a minimum number of incidents for a cluster. The algorithm then applies different (self-adjusting) search distances across the study setting. For example, very dense areas may have a higher number of small clusters while a more sparsely populated area would have fewer large clusters. By adjusting how "dense" is defined across the study area, HDBSCAN allows for clusters to be identified despite density being nonuniform across the study area.

1   Open the Density-based Clustering tool.

2   For Input Point Features, select Shotspotter Gunshots 2018.

3   For Output Features, type **Gunshots_HDB**.

4   For Clustering Method, select Self-Adjusting (HDBSCAN).

5   For Minimum Features Per Cluster, type **25**.

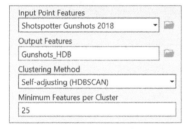

6   Run the tool.

The new Gunshots_HDB layer appears in the Contents pane.

7b

7   Turn off the Gunshots_DB_distance layer.

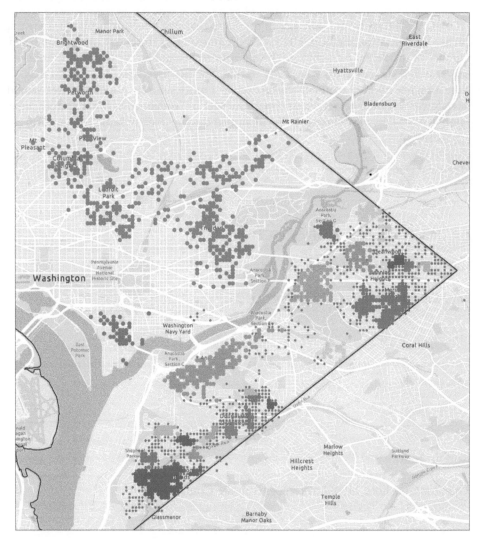

In total, 25 unique clusters of gunshot incidents were identified. To the east and south, a large number of smaller clusters appear, reflecting the high density of points in this area. Points to the north and northwest, mostly identified as noise by DBSCAN, are all now in a single cluster.

A Features Per Cluster chart is created by HDBSCAN, similar to DBSCAN. Unique to HDBSCAN is a chart displaying membership probability. This information can help you understand which points are central to the formation of the cluster.

8   In the Contents pane, double-click Distribution Of Membership Probability.

The display of this chart looks similar to the Features Per Cluster charts created previously. However, rather than presenting the number of cases across clusters, the Distribution Of Membership Probability chart provides a breakdown of the number of cases falling within each bin of probability scores. Simply put, the higher a point's probability score, the better the point fits into its assigned cluster. From the chart, you can see that 2,397 points have a probability score of between 0.97 and 1.00. These points can be considered the core of their respective clusters.

9   In the chart, click the 0.97 – 1.00 bin.

The corresponding points are now selected on the map.

Identify clusters with OPTICS

OPTICS is the final algorithm type used in density-based clustering. OPTICS provides an enhanced ability to adjust results to best reflect the contextual nature of clusters within a study area. For example, if you think that too many points have been assigned to a given cluster, you can adjust the parameters so that multiple clusters are instead created. OPTICS applies search distances by measuring the distance between each point and its closest neighbor. These distances are then placed into a reachability plot with a series of small distances representing clusters and large distances representing jumps from one cluster to another cluster (or a jump between a cluster and noise).

1   Open the Density-based Clustering tool.

2   For Input Point Features, select Shotspotter Gunshots 2018.

3   For Output Features, type **Gunshots_OP**.

4   For Clustering Method, select Multi-Scale (OPTICS).

5   For Minimum Features Per Cluster, type **25**.

You will leave the Search Distance field blank to allow OPTICS to identify an optimal distance.

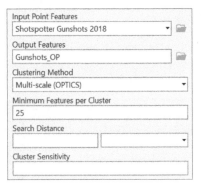

6   Run the tool.

The new Gunshots_OP layer appears in the Contents pane.

7   Turn off the Gunshots_HDB layer.

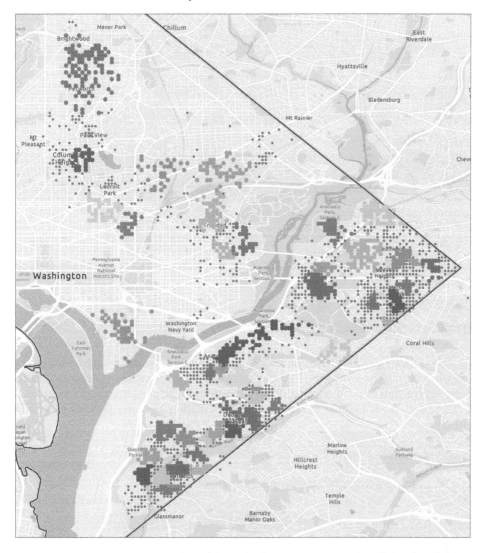

The OPTICS algorithm identified 44 individual clusters, substantially more than the DBSCAN and HDBSCAN processes. The Reachability Chart provides information on how distances between points were used in the creation of clusters.

7b

8 In the Contents pane, double-click Reachability Chart.

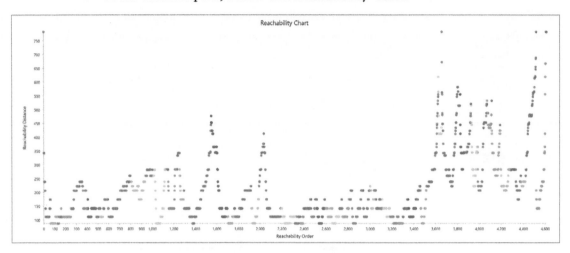

The Reachability Chart shows the distances between points, with peaks in distances representing a new cluster or noise.

Adjust clusters with cluster sensitivity

By default, OPTICS looks for large increases in point distances when distinguishing clusters from one another. You can specify a lower threshold for distinguishing between clusters by adjusting the cluster sensitivity. The cluster sensitivity ranges from 0 to 100, with higher numbers generally creating a greater number of smaller clusters.

1 Open the Density-based Clustering tool.

2 For Input Point Features, select Shotspotter Gunshots 2018.

3 For Output Features, type **Gunshots_OP_SEN**.

4 For Clustering Method, select Multi-Scale (OPTICS).

5 For Minimum Features Per Cluster, type **25**.

6 For Cluster Sensitivity, type **100**.

Input Point Features
Shotspotter Gunshots 2018
Output Features
Gunshots_OP_SEN
Clustering Method
Multi-scale (OPTICS)
Minimum Features per Cluster
25
Search Distance
Cluster Sensitivity
100

7 Run the tool.

The new Gunshots_OP_SEN layer appears in the Contents pane.

8 Turn off the Gunshots_OP layer.

The OPTICS algorithm identified 55 individual clusters.

9 Open the Reachability Chart.

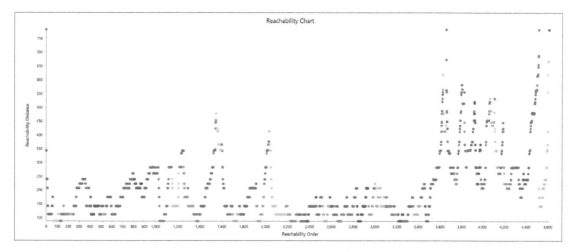

The Reachability Chart shows that new clusters were specified at smaller distances compared with the default OPTICS algorithm.

7b

Exercise 7c: Measure multivariate clustering

Crime analysis highlight 7c: Analyzing overlapping data in polygon features

More and more agencies are beginning to explore ways to evaluate the impact of patrol strategies and deployment plans in relation to identified hot spots. Chiefs, commanders, and others recognize the first step in the evaluation of any sort of plan is to be sure that resources were deployed as intended. From there, the goal should be to monitor the continued deployment, evaluate the impact over time, and then to adjust as necessary. Using GIS techniques, crime analysts can identify areas of their jurisdictions where officer activity highly overlaps with hot spots and, conversely, where officer activity and hot spots are more spatially diffused. Most often, crime hot spots are displayed on a map using the Kernel Density tool. It can sometimes be difficult, however, to then overlay police activity such as directed patrols or traffic stops, or other means of police presence. Creating a contiguous grid using tools in ArcGIS Pro (such as Create Fishnet or Create Thiessen Polygon) enables analysts to use grid cells as a means of further measuring data within defined and smaller areas. The incident count and spatial join tools can then be used to measure the number of features from various layers (for example, police activity and crime) falling within each cell. Analysts can then categorize grids as being high in crime, high in police activity, or high in both crime and police activity. The resulting grid can serve to prompt discussion about why some grid cell areas may get more incidents or more police attention than others, and allows a simple reference tool for adjustments.

—*Debra J. Piehl, DDACTS analytical specialist, International Association of Directors of Law Enforcement Standards and Training (IADLEST)*

In this exercise, you will use the Multivariate Clustering tool to identify how the different attribute fields are concentrated within the grids. Spatial clustering is often of interest to crime analysts, but feature characteristics are also important to consider. In this exercise, you will measure how the five fields contained within the grids layer are distributed across Santa Rosa, California. The Multivariate Clustering tool uses a K-means algorithm to identify clusters of feature attributes.

7c

Identify multivariate clusters

1 In the Catalog pane, in the Maps folder, click Multivariate to open the map for this exercise.

This map displays a layer of contiguous, 1,000-by-1,000-foot grids in Santa Rosa, California. The attribute table contains fields measuring activity occurring between 1/1/2020 and 7/14/2020. The number of security checks, extra patrols, and traffic stops conducted by officers is measured within each grid. Two additional fields measure the number of shots fired incidents and traffic accidents.

OBJECTID *	Shape *	Id	SecurityCheck	ExtraPatrol	TrafficStops	Shots_Fired	TrafficAccidents
1	Polygon	0	1	0	0	0	1
2	Polygon	0	0	0	0	0	0
3	Polygon	0	4	0	9	0	2
4	Polygon	0	0	0	0	0	0
5	Polygon	0	0	0	0	0	1

2 Search for and open the Multivariate Clustering (Spatial Statistics) tool.

3 In the Multivariate Clustering tool pane, for Input Features, select Santa Rosa Grids.

4 For Output Features, type **Grids_Clusters**.

5 For Analysis Fields, check the check boxes next to these fields:
- SecurityCheck
- ExtraPatrol
- TrafficStops
- Shots_Fired
- TrafficAccidents

6 For Clustering Method, select K Means.

> **Note:** K-medoids is the other available clustering method. K-medoids is more robust to noise and data outliers, such as a statistical median.

7c

7   For Initialization Method, select Optimized Seed Locations.

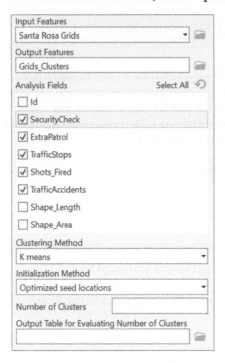

8   Run the tool.

The messages window of the output provides basic information about the importance of each variable in creating the attribute clusters.

9   Near the bottom of the Multivariate Clustering tool pane, click View Details.

Under Messages, the table provides summary statistics about each variable included in the K-means test. The R^2 value shows how influential each variable is in distinguishing clusters. In this example, traffic stops was the most influential variable, whereas shots fired was the least influential.

> **Messages**
>
> Start Time: Tuesday, April 13, 2021 9:05:20 AM
> Evaluating Optimal Number of Clusters....
>
> ⚠ WARNING 001326: The evaluation of the Optimal Number of Groups can take a significant amount of time.
>
> Optimal number of clusters is 2 based on the highest pseudo F-statistic.
> Variable-Wise Summary

Variable	Mean	Std. Dev.	Min	Max	R2
TRAFFICSTOPS	2.408131	6.235974	0.000000	57.000000	0.499434
TRAFFICACCIDENTS	0.988272	2.348593	0.000000	24.000000	0.417798
SECURITYCHECK	2.867866	7.230413	0.000000	101.000000	0.356191
EXTRAPATROL	0.191556	0.585243	0.000000	7.000000	0.241992
SHOTS_FIRED	0.222048	0.715591	0.000000	10.000000	0.147552

> Succeeded at Tuesday, April 13, 2021 9:05:37 AM (Elapsed Time: 17.29 seconds)

7c

> **Note:** You can access the messages of a previous run of the tool in the geoprocessing history. Search for the tool in the Geoprocessing pane, open the list next to the tool name, and click Open History.

The new Grids_Clusters layer appears in the Contents pane.

10 Turn off the Santa Rosa Grids layer.

The Multivariate Clustering tool found the five variables of interest clustered within two main groups.

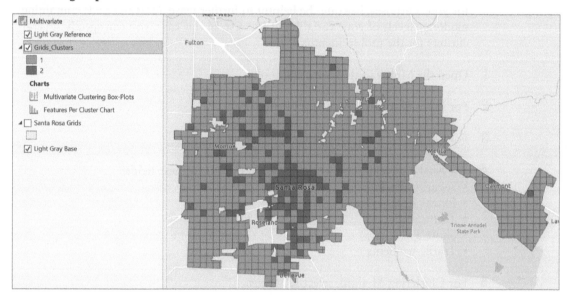

The symbology of the layer identifies the groups but does not communicate exactly how the attributes cluster within each group. You will explore the multivariate clustering box plots to glean this information.

11 In the Contents pane, double-click Multivariate Clustering Box-Plots.

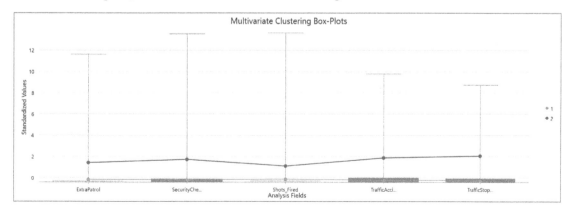

The box plot provides information on each variable's standardized value within the two clusters. In sum, the grids are separated into those that have low levels of each variable and those that have higher levels (between one and two standard deviations above the mean) of each variable.

Specify the number of multivariate clusters

It is important to consider the context of your data when interpreting multivariate clusters. In the current example, the grid features are small in size with a high frequency of grids with low or zero values for each variable. That probably explains why the K-means process identified only two broad categories of clusters. However, for your purposes, it would be helpful to further tease apart the grids containing moderate levels of your variables of interest. In this case, you can specify several clusters for the tool to identify.

1   Open the Multivariate Clustering tool.

2   For Input Features, select Santa Rosa Grids.

3   For Output features, type **Grids_Clusters_4**.

4   For Analysis Fields, check the check boxes next to these fields:
    - SecurityCheck
    - ExtraPatrol
    - TrafficStops
    - Shots_Fired
    - TrafficAccidents

5   For Clustering Method, select K Means.

6   For Initialization Method, select Optimized Seed Locations.

7c

7 For Number Of Clusters, type 4.

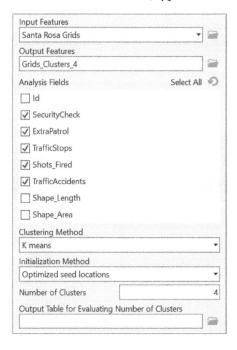

8 Run the tool.

The new Grids_Clusters_4 layer appears in the Contents pane.

9 Turn off the Grids_Clusters layer.

7c

Features are symbolized according to which of the four clusters the grid was assigned to. Because of the random component in finding seeds while using optimized seed locations for the initialization method, variations in clustering results will occur from one run of the tool to the next, and your results may not exactly match the one shown here. It is most appropriate to think of this tool as exploratory.

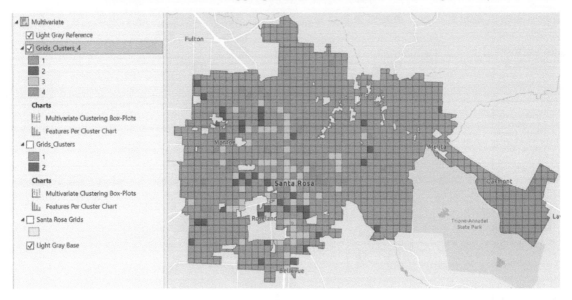

10 In the Contents pane, double-click Multivariate Clustering Box-Plots.

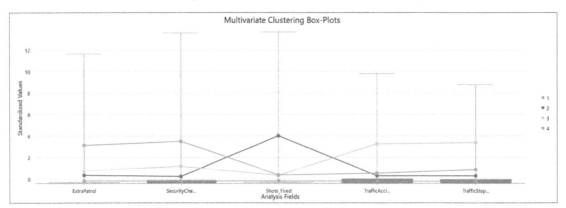

The box plot presents much more nuance than the prior example. By specifying four clusters, you have identified different levels of variables across clusters. Group 1 still consists of grids with low levels of all variables. Group 2 consists of grids with high levels of shots fired and low levels of all other variables. Group 3 consists of grids with higher levels of traffic accidents and traffic stops and moderate levels of all other variables. Finally, Group 4 consists of grids with high levels of extra patrol and security checks and lower levels of all other variables.

> **Note:** You can turn the clusters on or off in the box plots by clicking the cluster ID to the right of the chart.

Spatially constrain multivariate clusters

At its core, the K-means function underlying the multivariate clustering tool does not take space into consideration. Whereas the findings are plotted on a map to demonstrate how clusters are distributed across geography, features are assigned to clusters irrespective of spatial contiguity. As such, features belonging to the same cluster can often be far from one another. You can force the K-means function to identify spatially contiguous clusters by using the Spatially Constrained Multivariate Clustering tool.

1   Search for and open the Spatially Constrained Multivariate Clustering (Spatial Statistics) tool.

2   For Input Features, select Santa Rosa Grids.

3   For Output Features, type **Grids_Spatial**.

4   For Analysis Fields, check the check boxes next to these fields:
   - SecurityCheck
   - ExtraPatrol
   - TrafficStops
   - Shots_Fired
   - TrafficAccidents

5   For Cluster Size Constraints, select None.

6   For Number Of Clusters, type **4**.

7c

7   For Spatial Constraints, select Contiguity Edge Corners.

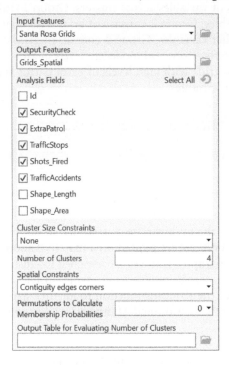

8   Run the tool.

The new Grids_Spatial layer appears in the Contents pane.

9   Turn off the Grids_Clusters_4 layer.

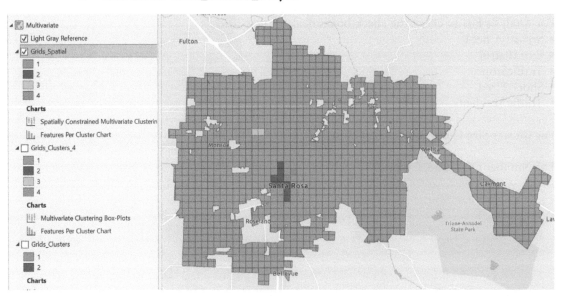

Because you forced clusters to be spatially contiguous, there is much less variability in the map, with the majority of features assigned to group 1. This is despite the K-means function successfully identifying four unique clusters in the data.

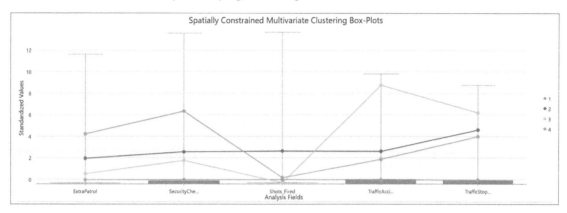

Exercise 7d: Perform generalized linear regression analysis

Crime analysis highlight 7d: Using regression analysis for place-based crime forecasting

Place-based crime forecasting techniques can have a great amount of value for police commanders. By understanding the environmental context of at-risk areas, police can better focus their resources and tailor strategies to address the place-based factors generating crime. Crime forecasting techniques can also help police understand whether their interventions are having the desired effects; conversely, areas unreceptive to specific interventions are also identified. This type of analysis typically involves statistical regression models that identify the factors significantly related to the outcome of interest (such as crime levels). In my dissertation research, I examined the impact of the metropolitan city of Bursa's (Turkey) citywide system and individual police-monitored CCTV cameras used to scan the landscape to address serious crime events. Using spatial analysis techniques, my research analyzed whether the environmental context (measured through risk-terrain modeling) impacted the deterrent effect of police-monitored CCTV cameras on aforementioned crime types. These results enable police to prioritize future CCTV sites according to the likelihood that surveillance cameras would generate crime reductions in the area.

—Emirhan Darcan, senior researcher, GLOBAL Policy and Strategy Institute

In this exercise, you will conduct a generalized linear regression analysis to diagnose the relative relationship that spatial risk factors and concentrated disadvantage have with street robbery in Denver's block groups.

Perform a Poisson regression analysis

1   In the Catalog pane, in the Maps folder, click Generalized Regression to open the map for this exercise.

This map displays census block groups in Denver, Colorado, symbolized by the number of street robbery incidents occurring in 2016.

The attribute table contains information on a range of sociodemographic and environmental characteristics for each block group. Six fields measure neighborhood factors associated with social disadvantage. These measures were standardized and summed in an index (z_index) reflecting the overall level of concentrated disadvantage in the block group compared with the average levels observed throughout Denver. An additional 12 fields measure the counts of environmental features that may operate as spatial risk factors for robbery. A risk terrain modeling analysis performed by Connealy and Piza (2019) tested the joint influence of these features on the occurrence of street robbery. For this exercise, the RTM data was aggregated and averaged across block groups to measure the aggregate neighborhood risk of crime (ANROC) value according to the process introduced by Drawve et al. (2016). The ANROC was standardized (z_anroc) to allow for easier comparison between the concentrated disadvantage and RTM data in the regression findings.

2   Search for and open the Generalized Linear Regression (Spatial Statistics Tools) tool.

3   For Input Features, select Block Groups Street Robbery.

4   For Dependent Variable, select Robbery.

When selecting the regression model type, you should understand the nature of the dependent variable. In the current example, robbery is measured as count of incidents, requiring the use of a count regression model.

5   For Model Type, select Count (Poisson).

6   For Explanatory Variable(s), select both z_index and z_anroc.

> **Note:** Additional variables could be added to the model by selecting other layers from the Explanatory Distance Features list. ArcGIS Pro automatically creates explanatory variables by calculating the distance from each of the explanatory distance features to the nearest input features. This is a helpful function when you do not have all variables of interest contained within a single feature layer.

7d

7   For Output Features, type **GLR_Robbery**.

8   Run the tool.

Unlike most geoprocessing tools, the primary output for generalized linear regression is not the feature layer added to the map. Rather, a text table displaying the findings of the regression analysis provides the primary information.

9   At the bottom of the tool pane, click View Details.

10   Expand Messages, and increase the width of the pane to better read the results.

Under Messages, the table reports the findings of the regression analysis. The coefficient shows the relationship of each independent variable to the dependent variable (street robbery count). In the current example, both Z_INDEX and Z_RRS are positive, which indicates that these measures are associated with heightened counts of street robbery. Poisson regression coefficients are reported in logged form, which can be interpreted as the change in the *log* of the dependent variable for every one-unit change in the independent variable. In this example, every one-unit change in z_index and z_anroc changes the *log* of robbery by 0.121264 and 0.359401, respectively. The incidence rate ratio (IRR) is a more straightforward way to communicate findings of count regression models. Although ArcGIS Pro does not report the IRR, it can be determined by calculating the exponential value of the coefficient. The IRR reflects the change in the dependent variable in terms of a percentage increase or decrease. In this example, the IRR for z_index (1.13) indicates that street robbery increases 13 percent with each one-unit increase of z_index. The IRR for z_anroc (1.43) indicates that street robbery increases 43 percent with each one-unit increase of z_anroc.

7d

> **Note:** The Generalized Linear Regression tool also gives you the option to save the regression coefficients as a stand-alone table in your project.

```
Variable Coefficient [a] StdError z-Statistic Probability [b]  VIF [c]
Intercept    -0.076244 0.048962   -1.557194      0.119425     --------
  Z_INDEX     0.121264 0.011292   10.739197      0.000000*    1.052011
  Z_ANROC     0.359401 0.015177   23.680704      0.000000*    1.052011
```

In addition to the table of regression findings, the Generalized Linear Regression tool creates a feature layer (GLR_Robbery) that appears in the Contents pane. The attribute table for the GLR Robbery layer contains the z_index and z_anroc values for each feature, as well as the observed dependent variable (robbery) count, predicted dependent variable count, and deviance residuals (also known as the random error term of the regression equation).

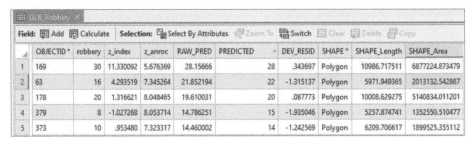

	OBJECTID *	robbery	z_index	z_anroc	RAW_PRED	PREDICTED	DEV_RESID	SHAPE *	SHAPE_Length	SHAPE_Area
1	169	30	11.330092	5.676369	28.15666	28	.343697	Polygon	10986.717511	6877224.873479
2	63	16	4.293519	7.345264	21.852194	22	-1.315137	Polygon	5971.949365	2013132.542887
3	178	20	1.316621	8.048465	19.610031	20	.087773	Polygon	10008.629275	5140834.011201
4	379	8	-1.027268	8.053714	14.786251	15	-1.935046	Polygon	5257.874741	1352550.510477
5	373	10	.953480	7.323317	14.460002	14	-1.242569	Polygon	6209.706617	1899525.355112

Exercise 7e: Perform geographically weighted regression

Research highlight 7e: Using geographically weighted regression to better understand local relationships

The spatial analysis of crime and the environmental correlates of hot spots have become popular areas of criminological research. This research has largely been concerned with how crime generators, crime attractors, and neighborhood demographic factors affect crime levels throughout a study area. However, statistical models employed in these studies have largely assumed that the relationships between environmental factors and crime is consistent across space. This ignores the possibility of spatial heterogeneity: the relationship between crime and a given environmental factor differing across space within an individual study setting (Andresen and Ha 2020; Wheeler and Steenbeek 2020). Geographically weighted regression has emerged as an analytic tool to model variable relationships that encompass such spatial heterogeneity. Rather than modeling relationships between variables for an entire area, geographically weighted regression constructs a separate model for every feature in a dataset. Through this function, geographically weighted regression can model spatially varying relationships that elude global regression modeling (Anselin 1988). Geographically weighted regression has been used to test the local effect of a wide range of factors on crime, including immigration (Andresen and Ha 2020; Graif and Sampson 2009), neighborhood diversity (Graif and Sampson 2009), concentrated disadvantage and collective efficacy (Becker 2019), and alcohol outlet density (Cameron et al. 2016).

In this exercise, you will perform a geographically weighted regression analysis to diagnose the local relationships between violent crime and the pertinent neighborhood measures.

Perform a continuous geographically weighted regression analysis

1 In the Catalog pane, in the Maps folder, click GWR to open the map for this exercise.

The map displays counties in Texas symbolized by violent crime rate. The attribute table contains fields measuring neighborhood health and well-being. All the data for the layer was collected from the 2018 County Health Rankings report.

2 Search for and open the Geographically Weighted Regression (GWR) tool.

3 For Input Features, select Texas Counties Health Rankings.

4 For Dependent Variable, select Violent Crime Rate.

As is the case with more traditional regression analysis, you should understand the nature of the dependent variable when conducting geographically weighted regression. In the current example, violent crime rate is reported as a continuous measure requiring the use of a continuous regression model.

5 For Model Type, select Continuous (Gaussian).

6 For Explanatory Variable(s), check the check boxes for Percent Fair/Poor Health, Percent Frequent Physical Distress, and Percent Frequent Mental Distress.

7 For Output Features, type **GWR_Violence**.

8 For Neighborhood Type, select Number Of Neighbors.

9 For Neighborhood Selection Method, select Golden Search.

> **Note:** The prior two selections enable the GWR model to adjust the number of neighborhoods for each feature based on an optimized distance. You can restrict these parameters by selecting fixed distances and intervals form the lists.

10 For Minimum Number Of Neighbors, type **35**.

This is the recommended minimum neighbor amount for the geographically weighted regression function in ArcGIS Pro.

11 Run the tool.

Check the model parameters

Before finalizing a geographically weighted regression analysis, it is important to check the model parameters and model fit statistics to determine whether any changes are warranted. This is especially true when using number of neighbors as the neighborhood type and golden search as the neighborhood selection method. This method runs a variety of models using a range of neighbors. Sometimes, the method may not report the model with the lowest Akaike information criterion (AIC). This suggests that a model using the number of neighbors corresponding to the lowest AIC may provide the better fit.

1 At the bottom of the Geographically Weighted Regression tool pane, click View Details.

The results table contains a warning indicating that the final model did not have the lowest AIC. You will determine which of the alternative models provides the best fit.

2 In the results table, expand Messages.

The golden search results indicate that a model with 104 neighbors achieved the lowest AIC value.

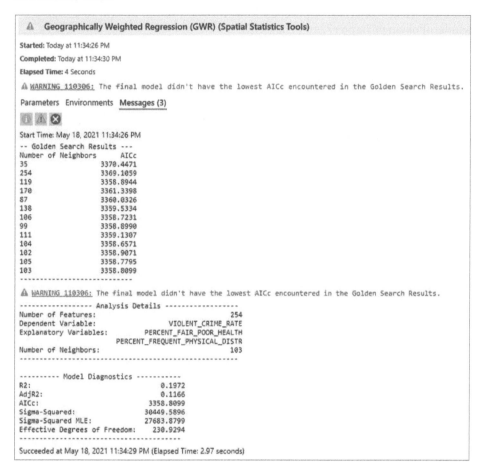

You will rerun the analysis with 104 neighbors specified.

Perform a geographically weighted regression analysis with number of neighbors specified

1 Open the Geographically Weighted Regression tool, if necessary.

> **Note:** If you closed the Geoprocessing pane already, repeat the steps from the "Perform a continuous geographically weighted regression analysis" section.

2 For Neighborhood Selection Method, select User Defined.

3 For Number Of Neighbors, type **104**.

4 Expand Additional Options.

5 For Coefficient Raster Workspace, click the browse button, browse to C:\
ModernPolicing\Chapter7, and click OK.

6 Run the tool.

> **Note:** Running the updated model automatically overwrites the original tool output (GWR_Violence).
> You do not need to remove this layer from the Contents pane to overwrite it.

The primary output of the Geographically Weighted Regression tool is similar to that of the Generalized Linear Regression tool: a feature layer symbolized by residual values and charts displaying relationships between variables, distribution of residuals, and a box plot of residuals and predicated probabilities. The differences between the regression approaches are evident from inspecting the GWR attribute table, which contains fields reflecting coefficients and standard errors across all features for each explanatory variable. This information enables the measurement of how relationships vary across space.

Visualize spatial relationships across space

There is currently no consensus about how to assess the confidence in coefficients from a geographically weighted regression. However, a popular approach to evaluate the coefficients is to divide the coefficients by the standard error provided for each feature. You will symbolize the GWR_Violence feature class in this manner for the Percent Frequent Mental Distress variable.

1   In the Contents pane, right-click GWR_Violence, and click Symbology.

2   For Primary Symbology, select Graduated Colors.

3   For Field, select Coefficient (PERCENT_FREQUENT_MENTAL_DISTRES).

4   For Normalization, select Std. Error (PERCENT_FREQUENT_MENTAL_DISTRES).

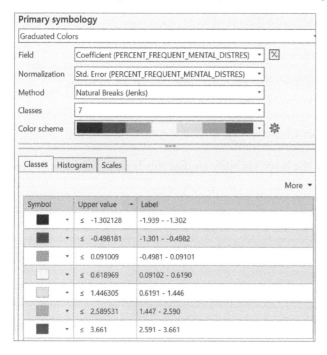

The map displays the normalized coefficients over space, with the largest coefficients appearing in the western portion of Texas.

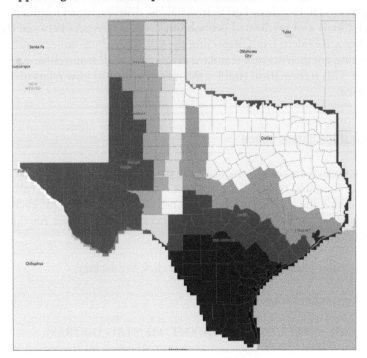

Another way to visualize the spatial relationships is through the coefficient raster layers created by the Geographically Weighted Regression tool. By default, the coefficient variables are created with stretch symbology. You will symbolize the raster layers so that they better communicate this spatial relationship.

5 In the Contents pane, right-click GWR_Violence_PERCENT_FREQUENT_ PHYSICAL_DISTR, and click Symbology.

6 For Primary Symbology, select Classify.

7 In the Contents pane, adjust the layer visibility so that this layer can be clearly seen.

Like the normalized feature layer, the raster layer shows that the effect of the percent frequent physical distress variable on violent crime rates is generally higher in the eastern counties and decreases as you travel west.

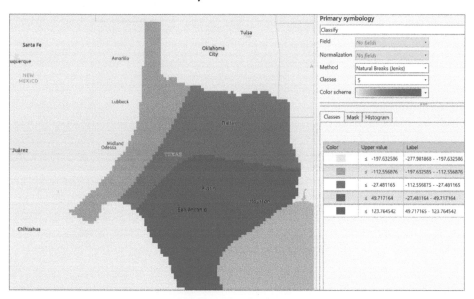

8 Repeat steps 1 through 7 for each of the other independent variables to show how they vary over space.

Summary

The tools and exercises covered in this chapter allow you to take spatial analysis to an advanced level. Designed with spatial analysis in mind, these tools and algorithms surpass the tools presented in previous chapters (effective though they may be). Being able to perform tasks such as measuring colocation of point features or performing various types of regression analyses increases your ability to identify spatial relationships in your data, which can greatly inform place-based policing interventions.

7e

Chapter 8
Automating crime analysis processes

Overview

In this chapter, you will learn how to perform automation in ArcGIS Pro. The automation tools include tasks that walk users through manually performing specific processes and models that create custom geoprocessing tools for completing processes.

You will acquire the following skills upon completing the exercises in this chapter:
- Creating task items and task processes
- Creating actions by recording mouse clicks
- Creating model diagrams in ModelBuilder™
- Editing existing models
- Creating and renaming model parameters
- Creating and renaming model variables

Download and install the data

Before working on the exercises, you will obtain the necessary data.

1 Go to www.arcgis.com and sign in with your ArcGIS Online account credentials.

2 Type **Modern Policing Using ArcGIS Pro** (**Esri Press**) in the search box, and click the Groups tab. Make sure that the Only Search In Your Organization option is turned off.

3 Click the Modern Policing Using ArcGIS Pro (Esri Press) group.

4 On the group heading, click Content.

5 Click the chapter 8 file and download it.

6 Locate the zip file you downloaded to your local drive, right-click, and extract it to C:\ModernPolicing.

This will create a folder named Chapter8 in the ModernPolicing folder.

This folder contains an ArcGIS Pro project and data you will use for the exercises in this chapter.

Exercise 8a: Create task items and processes

Tasks support common crime analysis functions by providing a set of preconfigured steps that guide users through a repeatable workflow. As an interactive guide, tasks help ensure the uniformity of work products across multiple analysts. This can be particularly helpful in training analysts who are newly hired or recently transferred into a new assignment.

Create a task

1   Open Chapter8.aprx in C:\ModernPolicing\Chapter8.

The project shows a map named Tasks that displays locations of aggravated assaults between January 1, 2020, and June 30, 2020, in Fayetteville, North Carolina. The Fayetteville Police Department (FPD) patrol sectors also appear on the map.

You will use this data to create a Task item that walks users through the process of counting the number of aggravated assault points within each patrol sector.

2   On the View tab, in the Windows group, click Tasks.

3   In the Tasks pane, click New Task Item.

The Task Designer tool opens in the Catalog pane. The General tab displays metadata for the processes in the task item. This information is helpful for communicating the goal of the task. You will populate these parameters.

4   For Name, type **Count incidents in sectors.**

5   For Author, type **Modern Policing.**

6   For Summary, type **Follow this task to create a new feature class that sums the number of crimes in each patrol sector.**

7   For Description, type **This task uses the following tools: 1) incident count, 2) add new field, and 3) calculate field.**

8   For Use Limitation, type **None.**

9   For Tags, type the following terms (press Enter after typing each term):
**incident count
sectors
new fields**

10 Check the Auto Increment check box.

The Auto Increment option automatically updates the version number whenever the task is updated.

You will create a task to guide the data extraction.

11 In the Tasks pane, click the New Task button.

12 In the Task Designer pane, for Name, type **Count incidents in sectors.**

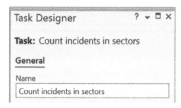

You are ready to add steps to the task.

Add a new step

1   In the Tasks pane, click the New Step button.

2   In the Task Designer pane, for Name, type **Incident Count**.

3   For Instructions, type **Click Run to start the step, and click Next Step to proceed to the next step.**

> **Note:** Text added in the ToolTip box appears when users pause over the task step. A ToolTip is not needed in this exercise given the self-explanatory name of the step.

4   In the Step Behavior section, select Manual.

> **Note:** The Manual option requires users to complete the current step manually and proceed to the next step. But step behavior also allows you to run a task automatically and proceed to the next step. You can also hide steps so that the process is not visible to users. These settings may be beneficial for processes that are run regularly or when users do not need to learn how to perform the tasks manually.

Create task steps by recording mouse actions

ArcGIS Pro allows users to create task steps by either recording their mouse clicks or manually adding tools to the task. You will create the first task steps by recording your mouse clicks.

1   In the Task Designer pane, click the Actions tab.

2 Below the description to Set The Command Or Geoprocessing Tool For The Step, click the Record button.

3 In the Crime Analysis gallery, click Summarize Incident Count.

Incident Count is applied as the geoprocessing tool in the task step. You will set the parameters of the tool.

4 In the Task Designer pane, click the Edit button to the right of Summarize Incident Count.

5 In the Command/Geoprocessing pane, for Type Of Command, select Geoprocessing Tool.

6 For Selected Geoprocessing Tool, click the folder to browse for and open the Summarize Incident Count tool.

7 In the Summarize Incident Count tool pane, for Input Features, select FPD Sectors.

8 For Input Summary Features, select Aggravated Assault 2020 (Jan. – Jun.)

9 For Output Feature Class, type **Sectors_AggAssault.**

8a

10 Keep Group Field empty, and click Done.

Create step 2 by manually adding tools

You will now finish the task by manually adding geoprocessing tools to the steps. For brevity's sake, you will not add text to instruction and description boxes. See the prior steps for guidance on populating these boxes.

1 In the Tasks pane, click the New Step button.

2 In the Task Designer pane, for Name, type **Add Field**.

3 On the Actions tab, click the Edit button.

4 In the Command/Geoprocessing pane, for Type Of Command, select Geoprocessing Tool.

5 Search for and add the Add Field (Data Management Tools) tool.

The parameters relate to the layer resulting from the first step of the task (which will not exist until the task is run). You will keep those parameters blank for now.

6 Click Done.

Create step 3 by manually adding tools

1 In the Tasks pane, click the New Step button.

2 In the Task Designer pane, for Name, type **Calculate Field**.

3 On the Actions tab, click the Edit button.

4 In the Command/Geoprocessing pane, for Type Of Command, select Geoprocessing Tool.

5 Search for and add the Calculate Field (Data Management Tools) tool.

6 Click Done, and close the Task Designer pane.

Run the task to count aggravated assaults in sectors

The three-step task item is complete and ready for use. You will run the task and add a field in the FPD Sectors feature class that denotes the number of aggravated assaults occurring from January through June 2020.

1 In the Tasks pane, click the arrow to the right of Count Incidents In Sectors.

Step 1 (1. Incident Count) of the task opens in the Tasks pane. You will run the step to sum the aggravated assault incidents across FPD patrol sectors.

2 Click Run.

8a

The new Sectors_AggAssault feature layer appears in the Contents pane. The sectors are symbolized according to the count of aggravated assaults that were intersected.

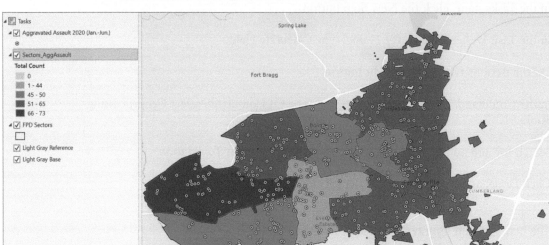

3 Open the Sectors_AggAssault attribute table.

The attribute table lists the aggravated assault count in the Total Count field.

Total Count is not a very intuitive name. You will use the remaining steps of the task to create a field named AggAssaults.

4 In the Tasks pane, click Next Step to move to the second step (2. Add Field).

5 For Input Table, select Sectors_AggAssault.

6 For Field Name, type **AggAssault**.

8a

7 For Field Type, select Long (Large Integer).

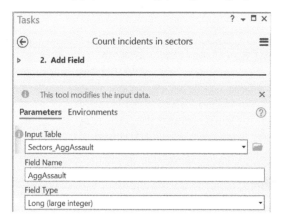

8 Click Run.

9 Click Next Step to move to step 3 (3. Calculate Field).

You will calculate the new AggAssault field with the values from the Total Count field.

10 For Input Table, select Sectors_AggAssault.

11 For Field Name (Existing Or New), select AggAssault.

12 For Expression Type, select Python 3.

13 In the Expression section, under Fields, double-click Total Count.

This builds the expression !TOTAL_CNT!

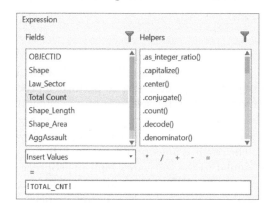

14 Run the tool, and click Finish.

In the attribute table, the AggAssault field is calculated to reflect the values of Total Count (in other words, the number of aggravated assaults occurring in the given sectors).

OBJECTID *	Shape *	Law_Sector	Total Count	Shape_Length	Shape_Area	AggAssault
1	Polygon	Alpha	63	222446.487457	383800992.878642	63
2	Polygon	AP	0	72742.689287	54380891.161452	0
3	Polygon	Bravo	50	63918.386716	130731212.341283	50
4	Polygon	Charlie	58	158503.590672	261645456.102246	58
5	Polygon	Delta	65	237175.095056	312842190.943415	65

Exercise 8b: Explore ModelBuilder

Crime analysis highlight 8b: Automating crime analysis workflows with ModelBuilder

Crime analysts are often tasked with monitoring specific locations in their jurisdictions, as well as generating data reports for personnel from the patrol level to command staff. These reports may include data pertaining to places of interest, specifically patrol beats, entertainment corridors, or transportation hubs. Because a major part of the analyst's workload involves mapping various types of data and performing spatial analysis, ModelBuilder can increase efficiency, potentially improve the accuracy of the data, and lead to more consistency in work products. ModelBuilder allows users to connect several geoprocessing tools, selections, and parameters, ultimately automating a workflow in which outputs feed into different tools until the desired process has concluded. Users can draw from the vast repository of GIS tools to create customized geoprocessing and use advanced programming languages for more complex models. As the analyst responsible for examining crimes that occur within a 500-foot buffer around a rail line that runs through the city, I use ModelBuilder to automate the steps used to build the buffer and pull and process the crime data from a geodatabase using established parameters. These models can be edited, shared, and run quickly, allowing current and future analysts to replicate the same process, and make changes to paths and data sources when necessary.

—*Carlena A. Orosco, police research & data analyst II, Strategic Planning, Analysis, and Research Center (SPARC), Tempe (Arizona) Police Department*

ModelBuilder allows you to create and save custom geoprocessing workflows within ArcGIS Pro. While tasks walk users through the steps necessary to perform a process, ModelBuilder enables a process to be completed with the selection of model parameters and the click of a button. ModelBuilder uses a visual programming language that allows users to drag tools into a workspace to create geoprocessing workflows. Manually writing code is not necessary. The result is a geoprocessing tool that runs in the ArcGIS Pro Geoprocessing pane, similar to the software's built-in geoprocessing tools. The ability to automate processes in ModelBuilder can be valuable to crime analysts because many common crime analyst tasks need to be repeated regularly.

This exercise introduces the ModelBuilder interface in preparation for subsequent exercises, which will involve creating models from scratch.

1 On the Analysis tab, in the Geoprocessing group, click ModelBuilder.

A new blank model opens. A contextual ModelBuilder tab also appears at the top of the project. You will use this tab to populate the (currently) empty ModelBuilder view to create a geoprocessing tool. The result will be a visual diagram that graphically displays each step in the geoprocessing tool.

Much of a model will be created using the tools contained within the Insert group. From here, you will add geoprocessing tools and variables to the model. You can also specify logic (for example, apply if-then statements) to your model and add labels communicating what each step of the model seeks to accomplish. A powerful function of ModelBuilder is the ability to set certain aspects of the model as parameters. Parameters generalize models by allowing you to change certain aspects as needed. For example, setting a crime feature class as a model parameter allows you to use a different layer in the geoprocessing function. In such a case, you can change the values of variables to suit their needs rather than modifying the structure of a model.

The View group allows you to adjust the visual layout of the diagram. As you begin to populate the model, it can become cluttered (especially when building complex models). The View tools enable you to automatically distribute the model components and adjust the visible scale to maximize visibility.

The Run group contains tools you will use to validate and run your model while in edit mode. You will use these tools to identify and correct errors as you build the model.

The Model group allows you to save and export the model as well as set the general properties and geoprocessing environment.

Under the ModelBuilder contextual tab, the Diagram tab gives you further control over the visual structure of the model. Here, you can edit the spacing between your model elements, rearrange the link connections between elements, and adjust the model layout and orientation.

New models are automatically saved in the toolbox in the active project folder and are accessible from the Catalog pane.

2 Close the ModelBuilder view.

With the basic structure of ModelBuilder explained, you will create custom geoprocessing tools in the remaining chapter exercises.

8b

Exercise 8c: Create a model to map incidents by x,y coordinates and append attributes from polygons

1   In the Catalog pane, in the Maps folder, click Model Map Incidents to open the map for this exercise.

The map displays police zones in Indianapolis, Indiana. A layer containing the boundaries of the police districts appears in the Contents pane with its visibility turned off. The project also contains two stand-alone tables containing residential burglary incidents and commercial burglary incidents occurring in 2019.

You will work with this data to create a model that first maps the tables using x,y coordinates and then appends attributes to the points from the polygon features. Before creating the model, you will familiarize yourself with the data layers.

2   In the Contents pane, right-click Burglary_Residental_2019, and click Open.

This table contains attribute fields denoting various incident characteristics, including the Uniform Crime Reporting (UCR) classification, UCR code, date and time of occurrence, case number, and address. Most pertinent to this exercise, the x,y coordinates appear to the far right of the table. You will use these fields to map the burglary incidents in the model.

Notably absent from the attribute table are any fields identifying the police command within which the incident occurred. You will append this data to the burglary points from the polygon feature layers.

3   Open the attribute table for Police Zones.

In the model, you will add the JURISDCTN and POLICEZONE attributes to each of the mapped burglary points.

4   Close the attribute tables.

Create a model

1   On the Analysis tab, in the Geoprocessing group, click ModelBuilder.

The ModelBuilder view opens. The model is saved in the current project toolbox (Chapter8.tbx) and is accessible from the Catalog pane. The Contents pane for Model Map Incidents, the active map, is visible. This will allow you to quickly add feature layers to the model by using a drag-and-drop process.

Before creating the model, you will change the name from *Model* to something that better reflects the process that will be automated.

2   On the ModelBuilder tab, in the Model group, click Properties.

3 In the tool properties, for Name, type **MapAppendPolygon**.

4 For Label, type **Map points & append polygon attributes**.

5 Ensure that the Store Tool With Relative Path option is checked.

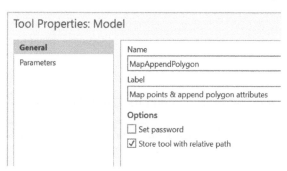

6 Click OK.

The Model view tab displays the name entered in the tool properties dialog box. However, the model file contained in Chapter8.tbx is still named *Model*. To change the name of the file, you must save the changes entered in the tool properties dialog box.

7 On the ModelBuilder tab, in the Model group, click Save.

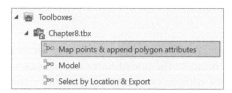

The model file has now been renamed.

Build an XY Table To Point process

1 On the Analysis tab, in the Geoprocessing group, click Tools.

2 In the search box, type **XY**.

3 From the search results, right-click XY Table To Point, and click Add To Model.

The XY Table To Point tool opens in the model as a gray rectangle, indicating that the tool is not ready to run. The tool will turn yellow after the tool properties are set. To the right of the tool is a gray oval that represents the output data that will be created by the XY Table To Point process.

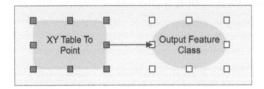

4 In the model view, right-click the gray XY Table To Point tool, and click Open.

5 In the XY Table To Point pane, for Input Table, select Burglary_Residential_2019.

6 For Output Feature Class, type **Residential_Burglary**.

7 For X Field, select X_COORD.

8 For Y Field, select Y_COORD.

> **Note:** The x,y coordinates may be automatically identified by ArcGIS Pro when no other fields in the attribute table are formatted as Double. You will not need to manually select the attributes in such cases.

It is important to remember the proper projection for the output layer when mapping points from coordinates. In the current example, the data should be projected in the Indiana State Plane coordinate system because the coordinates are x,y (as opposed to latitude and longitude). For the coordinate system value, you can either manually search for and assign NAD_1983_StatePlane_Indiana_East_FIPS_1301_ Feet as the coordinate system or assign the projection of another layer to the tool output. Because Police Zones is projected in Indiana State Plane, you will specify this layer in the pane.

9 For Coordinate System, select Police Zones.

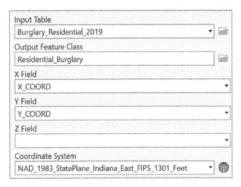

NAD_1983_StatePlane_Indiana_East_FIPS_1301_Feet is set as the coordinate system.

10 Click OK.

The Burglary_Residential_2019 table is added to the model. The arrow between the table and the tool shows the geoprocessing workflow. The output oval is now green, showing the output file (Residential_Burglary) that will be created.

> **Note:** The full names of the table and geoprocessing output are not entirely visible in the default view of the model. You can increase the size of the elements, and improve readability, by dragging the handles around each element.

You will now run the model to ensure the accuracy of the process before moving to the next portion of the model.

11 On the ModelBuilder tab, in the Run group, click Run.

A window appears that displays the progress of the model. The date and time of successful model completion is shown after the process concludes.

> **Note:** By default, the window remains open to allow for visual inspection of the process. You can check the Close On Completion check box so that the window automatically closes after the process runs.

Verify that the feature class was created correctly

The Residential_Burglary layer created by the process now appears in Chapter8.gdb. This layer can be added to the map from the Catalog pane. This allows for additional confirmation that the layer was created correctly.

1 In the Catalog pane, expand Chapter8.gdb to view its contents.

8c

2   Right-click Residential_Burglary, click Add To Current Map, and click the Model Map Incidents map tab.

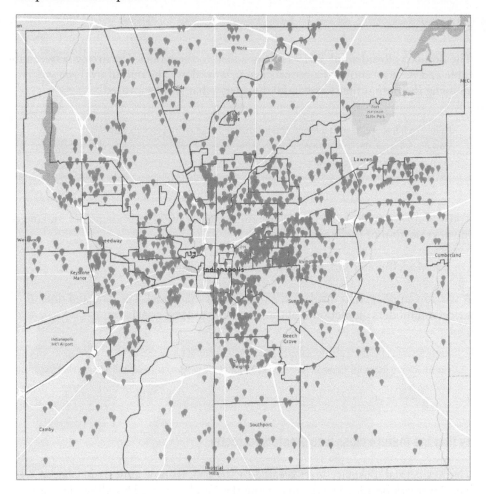

The Residential_Burglary layer appears in the Contents pane. Visible inspection of the layer confirms that the x,y coordinates were converted to points, as desired.

3   Return to the model view.

4   On the ModelBuilder tab, in the Model group, click Save to save the model.

Build an Append Attributes From Polygon process

With the residential burglary points mapped, you will create a process that appends the attributes from the Police Zones layer. Although you can set the model elements from within a tool, as in the prior step, this time you will drag a layer from the Contents pane.

1 In the Contents pane, click Police Zones, and drag it into the model view.

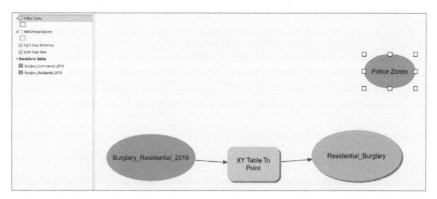

2 On the Analysis tab, in the Geoprocessing group, click Tools.

3 In the Geoprocessing pane, click the Toolboxes tab.

4 Open the Crime Analysis And Safety Tools toolbox.

5 Click the Join Attributes From Polygon tool, and drag it into the model view.

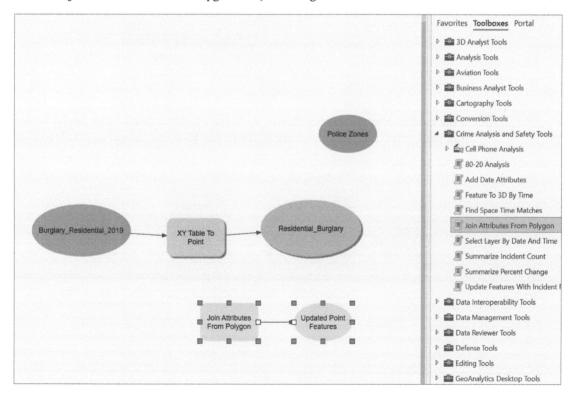

You now have all the necessary components in the ModelBuilder view to execute the tool. You will manually set the tool properties by connecting the necessary polygon and point layers.

6 In the model view, click Residential_Burglary. Holding down your mouse button, drag your pointer and the resulting arrow to the Join Attributes From Polygon tool.

7 Release your mouse button, and select Target Point Features from the pop-up menu.

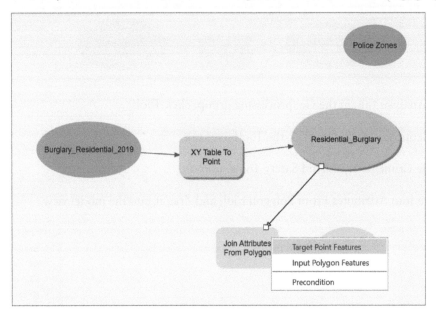

8 In the model view, click Police Zones. As before, holding down your mouse button, drag your pointer and the resulting arrow to the Join Attributes From Polygon tool.

9   Release your mouse button, and select Input Polygon Features from the pop-up menu.

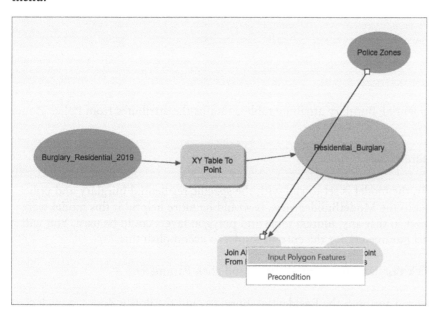

The model has become cluttered with all the tools and feature layers you have added. You can improve visibility by rearranging the model elements. ModelBuilder allows you to apply an automatic layout to your model.

10 On the ModelBuilder tab, in the View group, click Auto Layout.

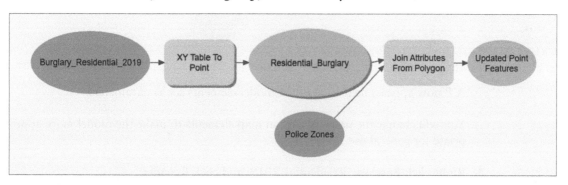

The model elements are now arranged in a manner that is easily viewed.

You will run the model to ensure the accuracy of the process before moving to the next step.

11 On the ModelBuilder tab, in the Run group, click Run.

The model runs successfully.

8c

12 Open the Residential_Burglary attribute table.

	DATE_	TIME	CASE	ADDRESS	X_COORD	Y_COORD	JURISDCTN	DISTRICT	POLICEZONE
FORCIBLE ENT-...	28-Mar-19	99:99	I190031605A	23 N DENNY ST	206433.64	1647239.44	IMPD	Northeast	NE26
FORCIBLE ENT-...	17-May-19	1:00	I190050706A	1444 PLEASANT ST ST	196624.87	1641868.22	IMPD	Southeast	SE15
FORCIBLE ENT-...	11-May-19	23:00	I190048912A	903 JEFFERSON AVE A...	199661.57	1650732.44	IMPD	Northeast	NE25
FORCIBLE ENT-...	2-Apr-19	5:00	I190032928A	2818 N N DEARBORN...	202930.05	1660789.41	IMPD	Northeast	NE10
FORCIBLE ENT-...	16-Apr-19	99:99	I190038835A	3630 N MERIDIAN ST	190290.13	1666180.67	IMPD	North	ND15

The Residential_Burglary attribute table contains the attributes from Police Zones.

Set model parameters

Currently, the model is set to run with the specific residential burglary and police zone data in the ModelBuilder view. It would be more helpful if this model were generalized so that any address table and polygon layers could be used. You will set the model parameters for the current model to accomplish this.

1 Right-click Burglary_Residential_2019, and click Parameter.

2 Repeat step 1 for both the Residential_Burglary and the Police Zones map elements.

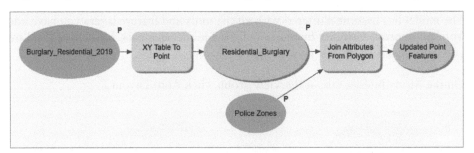

A *P* now appears near each of the model elements set as parameters.

You will change the name of certain map elements to make the model more appropriate for general use.

3 Right-click Burglary_Residential_2019, and click Rename.

4 In the text box, type **Address Table**.

5 Repeat these steps to change the name of model elements as follows:
  - Rename Residential_Burglary to **Output point layer**.
  - Rename Police Zones to **Polygon layer**.

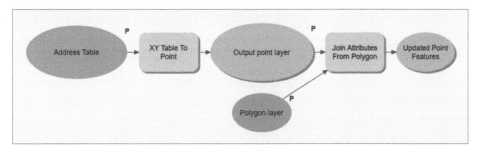

6 On the ModelBuilder tab, in the Model group, click Save.

7 Close the ModelBuilder view.

Run the final model

1 In the Catalog pane, open Chapter8.tbx.

2 Double-click Map Points & Append Polygon Attributes to open the model tool.

The model you created in the previous steps opens in the Geoprocessing pane as a model tool. You will run the model tool using the commercial burglary addresses table and the police districts polygons.

> **Note:** When the model tool first opens, a warning symbol appears next to the Output Point Layer text box because a file named Residential_Burglary already exists in Chapter8.gdb. If you run the model tool without changing the output layer name, this file will be overwritten.

3 For Address Table, select Burglary_Commercial_2019.

4 For Output Point Layer, type **Commercial_Burglary**.

8c

5 For Polygon Layer, select IMPDPoliceDistricts, and run the model tool.

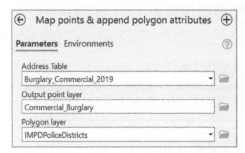

6 Open the Commercial_Burglary attribute table.

The x,y coordinates from the Burglary_Commercial_2019 table have been mapped, and the attributes from IMPDPoliceDistricts have been added to the attribute table.

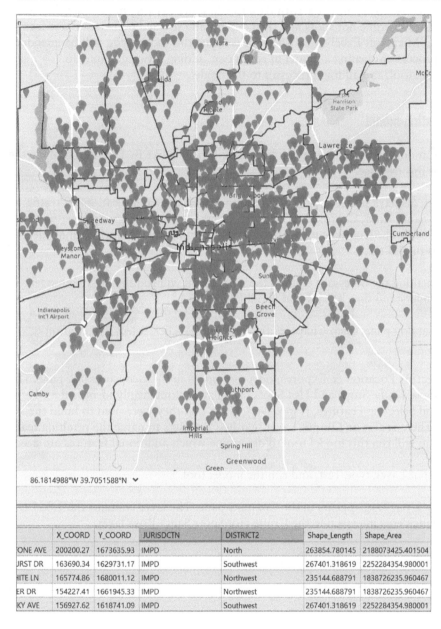

	X_COORD	Y_COORD	JURISDCTN	DISTRICT2	Shape_Length	Shape_Area
ONE AVE	200200.27	1673635.93	IMPD	North	263854.780145	2188073425.401504
URST DR	163690.34	1629731.17	IMPD	Southwest	267401.318619	2252284354.980001
HITE LN	165774.86	1680011.12	IMPD	Northwest	235144.688791	1838726235.960467
ER DR	154227.41	1661945.33	IMPD	Northwest	235144.688791	1838726235.960467
KY AVE	156927.62	1618741.09	IMPD	Southwest	267401.318619	2252284354.980001

8c

Exercise 8d: Edit a model to select incidents by location

In this exercise, you will work with data to edit a pre-existing model that first selects traffic stops in crime hot spots and then exports the selected points as a new feature class and Microsoft Excel spreadsheet. In creating this model, you will be introduced to the process of opening a model in edit mode, adding a variable function to a geoprocessing tool, and adding processes to the model.

Run the existing model

1 In the Catalog pane, in the Maps folder, click Model Select By Location to open the map for this exercise.

This map displays statistically significant crime hot spots and locations of traffic stops in Santa Rosa, California. These two layers are part of a group layer named Build Model. The Contents pane also contains a group layer named Run Model, which is turned off.

Before making the necessary edits to the model, you will run the existing model to better understand the changes that should be made.

2 In the Catalog pane, open Chapter8.tbx, and double-click Select By Location & Export.

The Select By Location & Export model tool opens in the Geoprocessing pane. The Santa Rosa Traffic Stops and Hot Spots layers are automatically identified as the Input and Selecting Features, respectively, given that they were used to build the model. There is also an Output Layer text box for users to name the resulting feature class. You will run this model tool to determine which additional features are needed.

3 For Output Layer, type **Test**, and run the model tool.

The new layer, Test, appears in the Contents pane. The layer contains the 1,355 traffic stop incidents intersecting the Crime Hot Spots Santa Rosa layer.

8d

4 Turn off the visibility for Traffic Stops Santa Rosa.

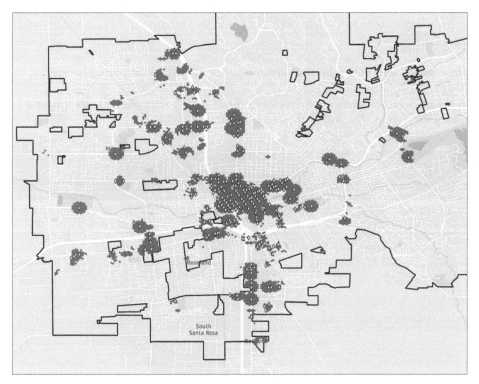

This will help you to see the Test layer better.

You will update this model to improve its general usability. You will make two edits. First, you will add search distance as a model variable. Although the current process is sufficient to answer questions related to the intersection of different features layers (for example, which points fall within hot spots?), it is not helpful to users interested in selecting features within a certain proximity of other features (for example, which points are within 1,000 feet of a hot spot?). The newly added variable allows the tool to answer both types of questions. Second, you will export the attribute table of the new output layer as a Microsoft Excel file. This format is helpful to disseminate incidents of interest to personnel who have no access to ArcGIS Pro.

8d

Add a model variable

1 In the Chapter8 toolbox, right-click the Select By Location & Export model, and click Edit.

The model diagram for this tool appears in the ModelBuilder view.

2 In the model, right-click the Select Layer By Location tool, click Create Variable, click From Parameter, and click Search Distance.

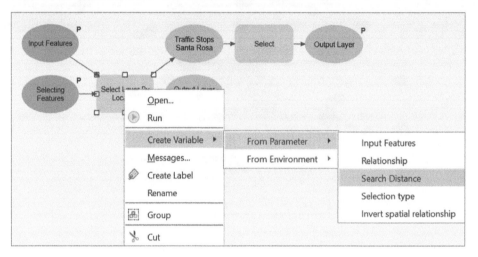

The Search Distance variable appears in the model. You must set the variable as a model parameter to enable users to specify a search distance when using this tool.

3 Right-click Search Distance, and click Parameter.

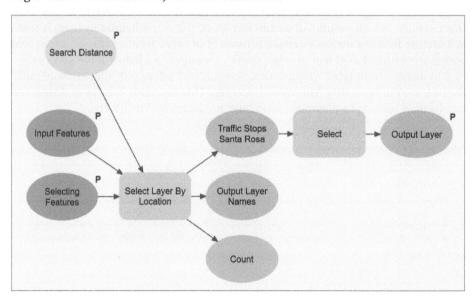

Build a Table To Excel process

You will adjust the model output to create a Microsoft Excel table in addition to a feature class. You will add the Table To Excel tool from the ModelBuilder tab.

1 In the model, in the Insert group, click Tools.

2 In the Tools search box, type **Table To Excel**.

3 In the search results, double-click Table To Excel to add it to the model.

The Table To Excel tool appears in the model. You will connect the tool to the output layer.

4 In the model, click Output Layer. Holding down your mouse button, drag your pointer to the Table To Excel tool.

5 Release the mouse button, and select Input Table from the pop-up menu.

6 In the model, right-click Output Excel File (.xls or .xlsx), and click Parameter.

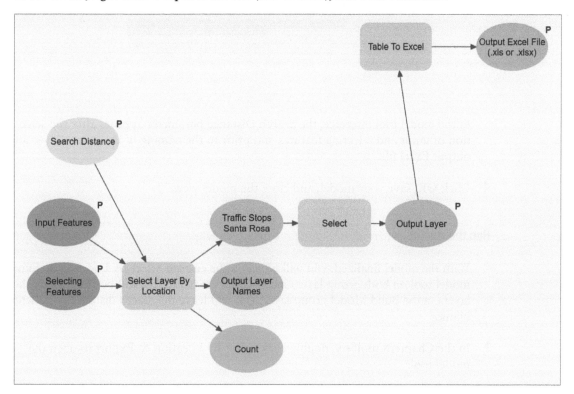

Reorder the model parameters

To finalize the model, you will reorder the parameters to adjust how they appear when users run the tool. Model parameters appear in the order in which they are set in the ModelBuilder window by default. You will adjust the order so that it better follows the decision-making process that users will follow. In particular, users will set the search distance variable prior to typing names for the output layer and Excel file.

1   On the ModelBuilder tab, in the Model group, click Properties.

2   In the Tool Properties dialog box, click Parameters.

3   Right-click the row selector to the left of Search Distance, and click Move Up.

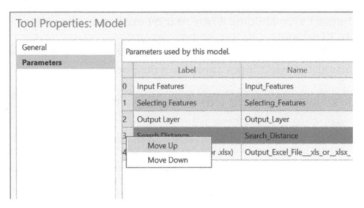

In the model tool interface, the Search Distance parameter appears after the selection of input and selecting features and prior to the naming of the output layer and output Excel file.

4   Click OK, save your model, and close the model view.

Run the model

With the model finalized, you will now run the custom Select By Location & Export model tool on both group layers in the Contents pane. You will begin by using the layers in the Build Model group layer. You will leave the search distance parameter blank.

1   In the Chapter8 toolbox, double-click Select By Location & Export to open the model tool.

2   For Input Features, select Traffic Stops Santa Rosa.

3   For Selecting Features, select Crime Hot Spots Santa Rosa.

4 For both Output Layer and Output Excel File (.xls or .xlsx), type **Stops_HotSpots**.

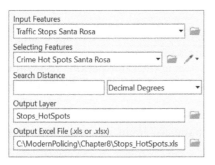

5 Run the tool.

6 Turn off the Test layer.

The new Stops_HotSpots layer, which includes the 1,355 traffic stops intersecting the crime hot spots, appears in the Contents pane. An Excel table containing these features also appears in C:\ModernPolicing\Chapter8.

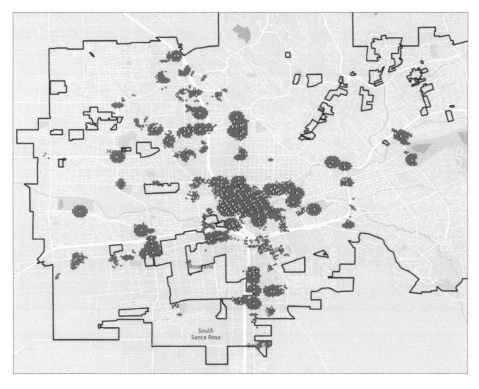

You will conclude this exercise by running the Select By Location & Export tool on the data in the Run Model group layer. You will prepare by adjusting the visibility of the layers in the Contents pane.

7 In the Contents pane, expand and turn on the Run Model group layer.

8d

8 Collapse and turn off the other layers, leaving the Santa_Rosa_Border and basemap layers on.

The Run Model group layer contains officer use-of-force events occurring from 1/1/20 to 7/5/20 as well as addresses receiving the highest amount of extra police patrols (identified with the 80/20 Analysis tool) during the same period. You will run the tool to identify use-of-force events occurring within 2,500 feet of the extra patrol addresses.

9 In the Select By Location & Export model tool pane, for Input Features, select Use of Force (1/1/20 – 7/5/20).

10 For Selecting Features, select Extra Patrol 80–20.

11 For Search Distance, type **2500**, and select Feet.

12 For both Output Layer and Output Excel File (.xls or .xlsx), type **UoF_Patrol**.

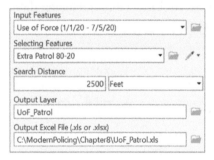

13 Run the tool.

14 Turn on the Use of Force (1/1/20 – 7/5/20) layer.

8d

The new UoF_Patrol layer, which includes the 21 use-of-force events occurring within 2,500 feet of the hot spot extra patrol addresses, appears in the Contents pane.

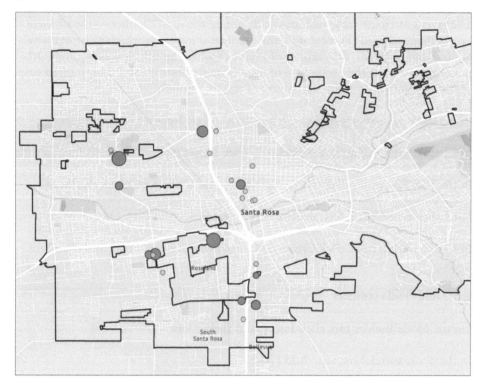

15 Browse to C:\ModernPolicing\Chapter8 to confirm the presence of an Excel table containing these features.

Exercise 8e: Create a model to create and calculate attribute fields

In this exercise, you will use ModelBuilder to build a tool that creates and calculates new fields in the attribute table. The model will include two processes. First, a new field will be added that calculates the number of events in which an injury did not occur. Second, a new field will be added that calculates the proportion of total force events that involved an injury. The field names will be set as a variable so that users can run the tool on either the suspect or the officer data.

Create a model from the Catalog pane

1 In the Catalog pane, in the Maps folder, click Model Attributes to open the map for this exercise.

8e

The map displays police zones in Indianapolis, Indiana, symbolized with graduated colors to show levels of police use of force (presented as standard deviations from the mean value) occurring between January and April 2020.

The layer's attribute table includes fields noting the total number of use-of-force events that occurred (Force_total), the number of events in which a suspect was injured (Sus_injured), the number of events in which an officer was injured (Off_injured), as well as the percentage of use-of-force events involving an injured suspect (Per_Sus_injured) or an injured officer (Per_Off_injured).

2   In the Catalog pane, right-click Chapter8.tbx, click New, and click Model.

3   In Chapter8.tbx, right-click Model, and click Properties.

4   In the Tool Properties dialog box, for Name, type **CreateCalculatePercentage**.

5   For Label, type **Create & Calculate Percentage Field**, and click OK.

6   In Chapter8.tbx, right-click Create & Calculate Percentage Field, and click Edit.

Build the Add Fields process

1   On the ModelBuilder tab, click Insert, and click Tools.

2   In the Tools search box, type **Add Fields**.

3   In the search results, double-click Add Fields (Multiple) to add it to the ModelBuilder view.

4   In the ModelBuilder view, double-click the Add Fields (Multiple) tool.

5   For Input Table, select IMPD Zones Use Of Force 2020 (Jan. – Apr.).

6   For Field Name, type **Sus_noinjury**.

7   For Field Type, select Long.

8   Click the Add Another button to add another field to the table.

9   For the second Field Name, type **Per_Sus_noinjury**.

10 For the second Field Type, select Long, and click OK.

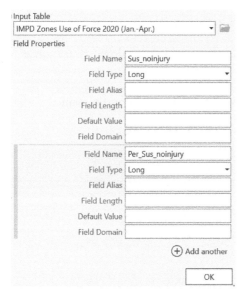

11 On the ModelBuilder tab, in the Run group, click Run.

The model runs and the Sus_noinjury and Per_Sus_noinjury fields appear in the attribute table.

12 Open the IMPD Zones Use Of Force 2020 (Jan.–Apr.) attribute table to confirm that the new fields have been created.

Rename processes, create variables, and set parameters

With the first portion of the model created, you will generalize the map processes to prepare the tool for use with any feature layers.

1 In the model, right-click the IMPD Zones Use of Force 2020 (Jan. – Apr.) input table, and click Rename.

2 In the text box, type **Input Table**.

3 In the model, right-click the Add Fields tool, click Create Variable, click From Parameter, and click Field Properties.

4 Right-click the Field Properties variable, and click Rename.

5 In the text box, type **Create Field Properties**.

6 For both Input Table and Create Field Properties, right-click, and select Parameter.

You will build a process to calculate the newly created Sus_noinjury field.

8e

Build the first Calculate Field process

1 Search for and add the Calculate Field tool to the model.

2 In the model, click the Updated Input Table process. Holding down your mouse button, drag the pointer to the Calculate Field tool.

3 Release your mouse button, and select Input Table from the pop-up menu.

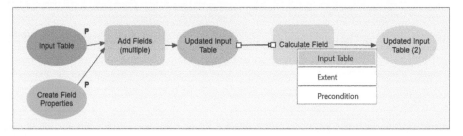

> **Note:** If the model layout gets cluttered, click the Auto Layout button to improve visibility.

4 In the model, double-click the Calculate Field tool.

5 For Field Name (Existing Or New), select Sus_noinjury.

6 For the Sus_noinjury expression, type **!Force_total! - !Sus_injured!**

7 Click OK.

8 Run the model to ensure that the process executes correctly.

The model runs, and the Sus_noinjury field is calculated.

9 Open the attribute table to confirm that the Sus_noinjury field contains values.

Sus_noinjury
0
2
142
10
12

Rename processes, create variables, and set parameters

1 In the model, right-click the Calculate Field tool, click Create Variable, click From Parameter, and click Field Name.

2 Right click the Calculate Field tool, click Create Variable, click From Parameter, and click Expression.

3 Right-click the Field Name variable, and click Rename.

4 In the text box, type **Difference Field**.

5 Right-click the Expression variable, and click Rename.

6 In the text box, type **Difference Expression**.

7 For both Difference Field and Difference Expression, right-click, and select Parameter.

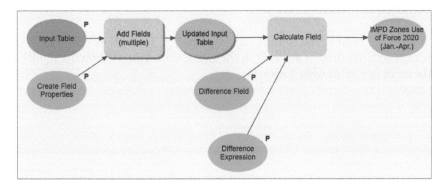

Build the second Calculate Field process

You will now build the process of the model that calculates the percentage of total use-of-force cases accounted for by noninjury cases.

1 Search for and add another Calculate Field tool to the model.

8e

2   In the model, click the output of the first Calculate Field tool, and connect an arrow to the Calculate Field (2) tool.

3   Select Input Table from the pop-up menu.

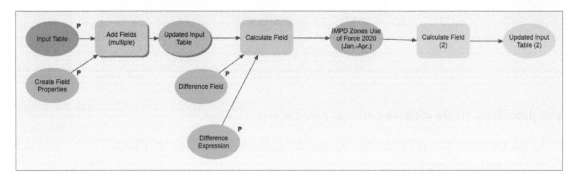

4   Double-click the Calculate Field (2) tool.

5   For Field Name, select Per_Sus_noinjury.

You will load an expression from your hard drive that calculates the percentage of use-of-force events not involving an injured suspect.

6   In the Calculate Field tool, near the bottom, click the Import button (folder icon) to import an expression.

7   Browse to C:\ModernPolicing\Chapter8\Expressions, and double-click Per_Sus_ noinjury.cal.

8   In the Calculate Field window, click OK.

The expression has been added to the Calculate Field (2) tool. However, as you will see in the next section, there is an error in the expression. You will run the model and correct the error before moving forward.

Update and export an expression

1 Run the model.

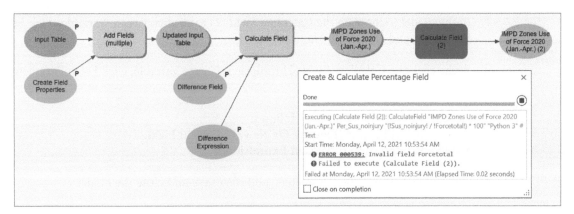

The progress window shows that the model failed to fully execute. The Calculate Field (2) tool is red in the model, which further signifies that the run was unsuccessful. The error message states that an invalid field (Forcetotal) was included in the expression. This field is missing an underscore, which you will add before saving the new expression.

2 In the model, double-click the Calculate Field (2) tool.

3 In the expression window, change !Forcetotal! to **!Force_total!**.

4 Click the green arrow to export the expression.

5 Browse to C:\ModernPolicing\Chapter8\Expressions.

6 For Name, type **Per_sus_noinjury_NEW**.

7 Click Save.

8 In the Calculate Field window, click OK.

9 Run the model again.

Sus_noinjury	Per_Sus_noinjury
0	<Null>
2	100
142	59
10	31
12	100

The model has successfully run and the Per_Sus_noinjury field now has values in the attribute table.

8e

Rename processes, create variables, and set parameters

You will set the variables and parameters for this process.

1 In the model, right-click the Calculate Field (2) tool, click Create Variable, click From Parameter, and click Field Name.

2 Right-click the Calculate Field (2) tool, click Create Variable, click From Parameter, and click Expression.

3 Rename the model variables as follows:
   - Rename Field Name (Existing Or New) to **Percent Field**.
   - Rename Expression to **Percent Expression**.

4 Set both variables to be Parameters, and then save and close the model.

Run the final model

With the model created, you will now run the model tool. Since the suspect fields were generated during the model creation, you will adjust the parameters so the officer noninjury difference and percentage fields are calculated.

1 In the Catalog pane, open Chapter8.tbx.

2 Double-click Create & Calculate Percentages to open the tool.

3 For the first Field Name, type **Off_noinjury**.

4 For the second Field Name, type **Per_Off_noinjury**.

5 For Difference Field, select Off_noinjury.

6 In the Off_noinjury expression window, type **!Force_total! - !Off_injured!**.

> **Note:** You can update the expression by either manually typing in the text box or adding fields to the expression from the Fields window.

7 For Percent Field, select Per_Off_noinjury.

8e

8 In the Off_noinjury expression window, type (!Off_noinjury! / !Force_total!) * 100.

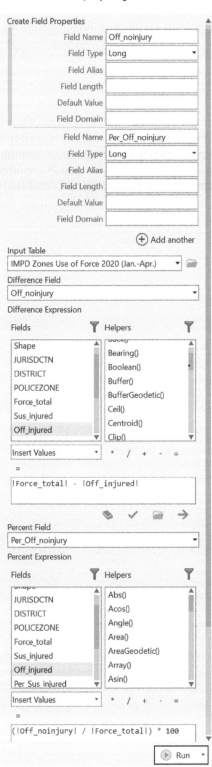

9 Run the tool.

The count and percentage of use-of-force events not involving injured officers are added to the attribute table.

Summary

Taking a step back from introducing specific analysis tools and techniques, this chapter focused on another important aspect of crime analysis: automating workflows. As demonstrated in the previous seven chapters, GIS requires a significant amount of data and manipulation of that data to perform key tasks, many of which are often repeated. This chapter introduced key concepts about customizing tasks and building models in ArcGIS Pro to enhance your efficiency and achieve a reliable, customizable degree of automation.

Chapter 9
Sharing your work

Overview

In this chapter, you will learn how to share your crime analysis products and data in a variety of ways. ArcGIS Pro allows users to share information as local files or within online data portals. You can customize each element of your products and data regardless of how information is shared.

You will acquire the following skills upon completing the exercises in this chapter:
- Creating layer packages
- Customizing feature labels
- Creating charts
- Editing chart properties
- Using the layout view
- Saving web maps in ArcGIS Online portals

Download and install the data

Before working on the exercises, you will obtain the necessary data.

1   Go to www.arcgis.com and sign in with your ArcGIS Online account credentials.

2   Type **Modern Policing Using ArcGIS Pro (Esri Press)** in the search box, and click the Groups tab. Make sure that the Only Search In Your Organization option is turned off.

3   Click the Modern Policing Using ArcGIS Pro (Esri Press) group.

4   On the group heading, click Content.

5   Click the chapter 9 file and download it.

6   Locate the zip file you downloaded to your local drive, right-click, and extract it to C:\ModernPolicing.

This will create a folder named Chapter9 in the ModernPolicing folder.

This folder contains an ArcGIS Pro project and data you will use for the exercises in this chapter.

> **Crime analysis highlight 9: Disseminating crime analysis findings**
>
> Evidence-based policing requires strategic decision-making at every level of the agency. A steady flow of information and data comes in at all times, which then needs to be processed and immediately disseminated into the field. As an NYPD crime analyst assigned to one of the city's patrol boroughs, my responsibility is to disseminate information and my findings to various levels of decision-makers such as supervisors, precinct commanders, borough-level chiefs, and chiefs at headquarters. My findings are quickly disseminated in the form of patrol-level alerts, identifying areas of concern that officers on the beat need to be aware of while on patrol. The alert is an identified geographic area experiencing a specific crime type spike. Details such as precinct, sector, platoon, offender m.o., victim information, and time and date are examples of details that get distributed. Such knowledge is invaluable to strategic planning, resource allocation, and connecting larger swaths of geographic areas that may involve other precincts or connect to the city as a whole. Eventually, successful strategies become a focal point of learning for commanders at CompStat meetings. The day-to-day functions of the NYPD show that crime analysis products are most useful when put into the hands of decision-makers.
>
> —*Tony Campos, borough crime analyst, New York City Police Department*

Exercise 9a: Create layer packages

Layer packages have the benefit of saving all aspects of data layers, including symbology, labeling, and field properties. Layer packages also include the dataset referenced by the map layers, meaning that others can work with your data without having access to the raw data files.

Symbolize feature classes

You will begin this exercise by symbolizing the property crime and sector layers. Features in the property crime layer will be symbolized according to their UCR crime type.

1   Open Chapter9.aprx in C:\ModernPolicing\Chapter9.

The project shows a map named Layer Package. The map displays locations of property crime incidents occurring during July 2020 in Fayetteville, North Carolina. The Fayetteville Police Department's patrol sectors also appear on the map.

The attribute table for the property crimes and patrol sectors contains fields noting the Uniform Crime Reporting (UCR) category and total number of property crime incidents reported in the sector (Prop_Crime), respectively. You will use the following data to create a layer package that can be shared with other ArcGIS Pro users.

2 In the Contents pane, click Property Crime (July 2020).

3 On the Appearance tab, in the Drawing group, click Symbology, and click Unique Values.

4 For Field 1, select UCR.

5 Symbolize each UCR category with a unique values symbology of your choice.

> **Note:** See chapter 1, exercise 1d for detailed instructions on symbolizing points with unique values.

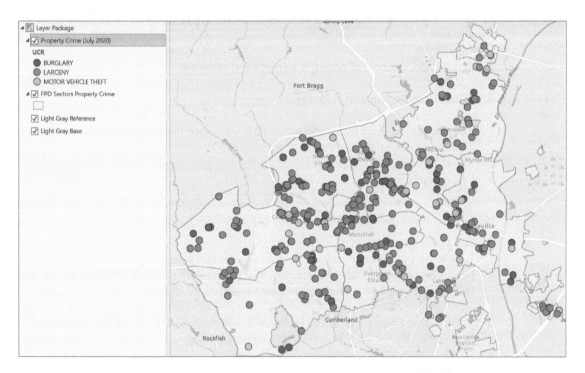

You will symbolize the patrol sectors by graduated colors according to their property crime level.

6 In the Contents pane, click FPD Sectors Property Crime.

7 In the Symbology pane, for Primary Symbology, select Graduated Colors.

8 For Field, select Prop_Crime.

> **Note:** Activate the Symbology pane from the Appearance tab, Drawing group if you previously closed it.

9a

9 For Color Scheme, select the Cividis color ramp.

Label the police sectors

As a final step prior to creating the layer package, you will label the police sectors so the total property crime count is visible on the map. You will load a premade expression to quickly populate the map.

1 In the Contents pane, right-click FPD Sectors Property Crime, and click Labeling Properties.

2 In the Label Class pane, click the Import button near the bottom of the Expression window.

3 Browse to C:\ModernPolicing\Chapter9\Expressions.

4 Click Prop_N.lxp, and click OK.

The expression is added to the pane. This will place the property crime totals on top of the sectors in the map view.

5 Click Apply.

Labels are placed over each of the patrol sectors. However, you will make changes before finalizing the labels. First, you will remove the multiple labels from the map so that each sector has a single corresponding label.

6 In the Label Class pane, click the Position tab, and click the Conflict Resolution button.

7 For Remove Duplicate Labels, select Remove All.

You will improve the visibility of the labels by increasing the size of the text.

8 In the Label Class pane, on the Symbol tab, click the General button.

9 Expand Appearance, and for Size, select 12 pt.

10 Expand Halo.

11 For both Outline Width and Halo Size, select 1 pt, and click Apply.

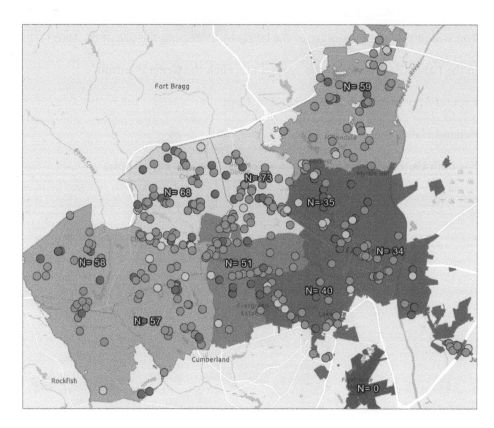

The symbology and label scheme of the map's layers are now finalized. You will create the layer package.

Save the feature classes as a layer package

All layers that will be included in the layer package must first be grouped together in the Contents pane.

1   In the Contents pane, select both Property Crime (July 2020) and FPD Sectors Property Crime, right-click, and click Group.

Property Crime (July 2020) and FPD Sectors Property Crime are now included in New Group Layer. You will change the group layer name.

2   In the Contents pane, double-click New Group Layer.

3   For Name, type **Property Crime: July 2020**, and click OK.

4   In the Contents pane, right-click the group layer, click Sharing, and click Share As Layer Package.

5   In the Package Layers pane, select Save Package To File.

> **Note:** You can also save layer packages to an ArcGIS Online account. Saving layer packages online follows the same procedure as creating web maps, which is covered in exercise 9d.

6 Under Item Details, for Name, click the Browse button.

7 Browse to C:\ModernPolicing\Chapter9 and save the layer package as **FPD_Prop_0720**.

8 For Summary, type **This layer package contains property crime points and sector totals for July 2020.**

9 For Tags, type the following terms, using commas to separate them: **property crime, sectors, Fayetteville**

10 In the Package Layers pane, click Analyze to check the layer package for errors.

11 Click Package.

The layer package is finalized. Whenever the layer package is loaded into an ArcGIS Pro project, it will maintain the symbology and transparency you applied to it.

Exercise 9b: Create and export charts

Add a chart to your map project

1 In the Catalog pane, in the Maps folder, click Charts to open the map for this exercise.

This map displays neighborhoods in Denver, Colorado, symbolized by overall crime rate. The Contents pane also includes a layer named Street Robbery Denver 2016, which is currently turned off.

9b

You will work with both layers to create charts in this exercise. You will first work with the Crime Rates By Neighborhood 2016 layer. The chart for this layer will display the distribution of crime rates across neighborhoods and as measures of central tendency (in other words, mean, median, and standard deviation). You will make formatting edits to the chart before saving it as an external file that can be shared.

2 In the Contents pane, click Crime Rates By Neighborhood 2016.

3 On the Data tab, in the Visualize group, click Create Chart, and click Histogram.

Histogram 1 appears in the Contents pane. You will add variables to create the histogram.

4 Open Histogram 1, if it is not already open.

5 In the Chart Properties pane, for Number, select Crime Rate Per 10000 People.

6 Under Statistics, check the check boxes for Mean, Median, and Std. Dev.

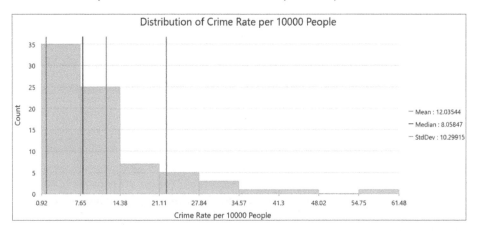

The chart has now been created and populated with the relevant variables.

The chart displays bins of crime rates across the x-axis and the counts of neighborhoods on the y-axis. Lines corresponding to the mean (12.03), median (8.05), and standard deviation (10.29) distances above and below the mean appear on the chart. The statistics header provides additional information on the data, showing the presence of 78 features (in other words, neighborhoods) in the layer and that crime rates range from 0.92 per 10,000 people to 61.48 per 10,000 people.

The chart properties are interactive, with the fields displayed in the map frame. To see the features that correspond with the different portions of the chart, select a data bin in the chart.

7 In the chart, click the 61.48 bin.

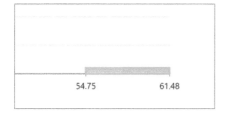

The neighborhood corresponding to this crime rate is selected on the map.

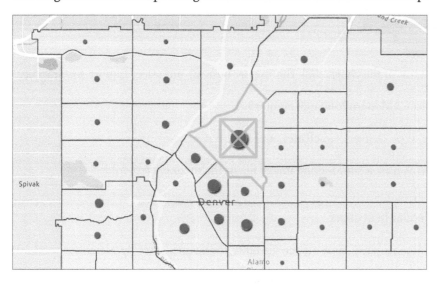

Adjust the chart properties and export the chart

You will adjust the title and other properties of the map before saving the chart as a stand-alone file.

1 In the Chart Properties pane, click the General tab (widen the pane if needed).

2 For Chart Title, type **Denver Neighborhood Crime Rates (2016)**.

3 For Legend Title, type **Central Tendency**.

9b

4   Keep all the other default values.

5   Near the top of the chart, click the Export button, and select Export As Graphic.

6   Browse to C:\ModernPolicing\Chapter9.

7   For Save, type **DenverCrimeRates**, and click Save.

The chart is now a stand-alone image file in the main project folder.

Create a calendar heat chart

1   In the Contents pane, turn off the visibility for Crime Rates By Neighborhood 2016.

2   Turn on the visibility for Street Robbery Denver 2016.

This layer displays the locations of street robbery incidents occurring throughout Denver. An important consideration is whether any temporal patterns are present in the data. Calendar heat charts can assist analysts in exploring such questions. Calendar heat charts visualize temporal patterns by aggregating incidents into a calendar grid. In this exercise, you will create a calendar heat chart to visualize how street robbery incidents fluctuated throughout the year. The result will be a heat chart in which each row corresponds to a month and each column corresponds to a day of the month.

> **Note:** If your data contains time of day, you can create a calendar heat map with a week view, in which each row corresponds to a day of the week and each column corresponds to an hour of the day.

3   In the Contents pane, click Street Robbery Denver 2016.

4   On the Data tab, in the Visualize group, click Create Chart, and click Calendar Heat Chart.

5   In the Chart Properties pane, for Date, select REPORTED_D.

6   For Type, select Year By Month And Day Of Month.

7   For Color Scheme, select Yellow-Green-Blue (Continuous).

8   In the Chart Properties pane, click the General tab.

9   For Chart Title, type **Street Robbery by Month and Day.**

10 For X Axis Title, type **Day.**

11 For Y Axis Title, type **Month.**

12 For Legend Title, type **Number of incidents.**

9b

The calendar heat map displays the temporal distribution of street robberies throughout the year. The shading shows robbery incidents generally concentrated in the latter portion (from the 25th onward) of July and August.

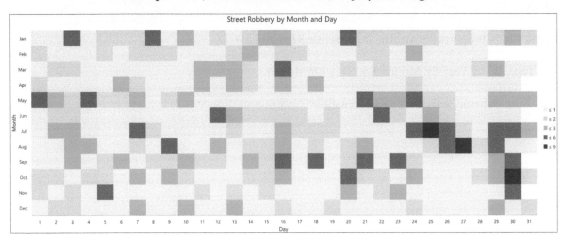

You can follow the previous steps to export the calendar heat chart as a stand-alone file.

Exercise 9c: Use the layout view to create and export maps

It is best to think of the layout view as the final area where a map is prepared for publishing. This preparation includes adding a legend, a scale bar, and other elements to the map to create a finished product. As with other aspects of ArcGIS Pro, a variety of options can be explored in layout view. There is also no limit to the number of separate layouts you can create. The final product you create in this exercise will be exported to Adobe Acrobat PDF, a file format that crime analysts use to share their products.

Create a layout

1   In the Catalog pane, in the Maps folder, click Layout to open the map for this exercise.

This map displays contiguous grid cells in Denver, Colorado. This layer resulted from risk-terrain modeling (RTM) and hot spots analyses meant to identify the micro-places most at risk of experiencing street robbery in 2016. Features are symbolized according to the method by which they were designated high risk: RTM, Hot Spot, both RTM & Hot Spot, or Neither.

You will create and export a map using this data in the layout view.

Layouts are added to your map project in a similar fashion as other map elements. An important consideration with map layouts is the dimension of the layout view.

Given the general shape of the Denver city boundary, you will select a landscape layout for your map.

2   On the Insert tab, click New Layout.

3   Under ANSI – Landscape, select Letter 8.5" x 11".

A new layout is added to the ArcGIS Pro project. The blank layout opens to the right of the Map tab. A Contents pane corresponding to the layout appears to the left.

4   On the Insert tab, click Map Frame.

The Map Frame list includes all maps and bookmarks contained in the current project.

5   Under the Layout header, select Denver.

6   To place the map in the layout, drag your pointer to draw a box that fills the entire layout view area where you want the map to appear, and release your mouse button.

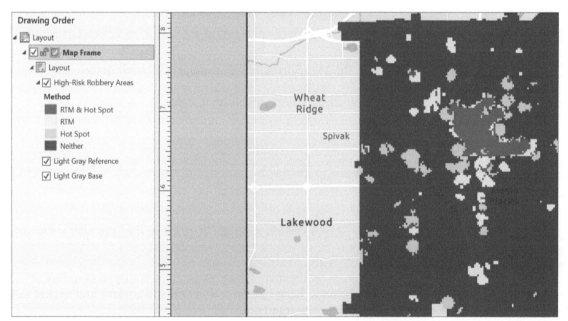

This effectively "pastes" the map into the layout view. The Contents pane is populated with the feature layers from the layout map view.

At this point, maps are not immediately accessible for editing (for example, for panning or zooming) in the layout view. Maps must first be activated within the layout view before you can adjust the scale. In the current exercise, you will pan the map downward so that the southern border of Denver is closer to the bottom of the layout view.

9c

7   Right-click anywhere in the layout view, and click Activate.

The Layout tab is replaced by the Map tab. You now have access to all the navigation tools to adjust the scale of the map.

8   On the Map tab, in the Navigate group, click the Explore tool.

9   Pan the map downward to move the southern border to the bottom edge of the layout.

With the map positioned to your liking, you will deactivate the map frame.

10   On the Layout tab, in the Map group, click Close Activation.

> **Note:** You can also activate the map from this same tab.

You will add necessary components to the map.

Add a north arrow

1   On the Insert tab, in the Map Surroundings group, click North Arrow.

2   Under North Arrows, select the ArcGIS North 1 style.

3   Drag your pointer to draw a custom-sized north arrow in the upper-left corner of the map.

The north arrow appears on the map. However, given its transparent nature, it blends in a bit with the underlying basemap. You will add a frame and background color to make the north arrow stand out.

4   In the Contents pane, double-click North Arrow.

5   In the Format North Arrow pane, click the Display button.

6   For Border, set the Symbol color to black.

7   For Background, set the Symbol color to white.

9c

8   For both Border and Background, set the X Gap and Y Gap to 0.1 in.

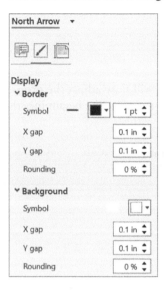

The north arrow now stands out more clearly from the underlying basemap.

Insert a scale bar

1   On the Insert tab, in the Map Surroundings group, click Scale Bar.

2   Under Scale Bar, select the Scale Line 1 style.

3   Drag your pointer to draw a custom-sized scale bar in the lower-right corner of the map.

4   In the Contents pane, double-click Scale Bar.

5   In the Format Scale Bar pane, click the Display button.

6   Under Border, set the Symbol color to black.

7   Under Background, set the Symbol color to white.

9c

8 For both Border and Background, set the X Gap and Y Gap to 0.1 in.

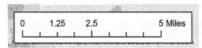

The scale bar lists distance intervals at four points. You will now adjust the properties so that distance intervals are only displayed three times: at the beginning, the middle, and the end of the scale bar.

9 In the Format Scale Bar pane, click the Properties button.

10 Under Numbers, for Frequency, select Divisions.

The scale bar now lists only three distance intervals.

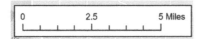

Insert a legend

1 On the Insert tab, in the Map Surrounds group, click Legend.

2 Drag your pointer to draw a custom-sized legend in the lower-right corner of the map, above the scale bar.

3 In the Contents pane, double-click Legend.

4 In the Format Legend pane, click the Display button.

5 Under Border, set the Symbol color to black.

6 Under Background, set the Symbol color to white.

7 For both Border and Background, set X Gap and Y Gap to 0.1 in.

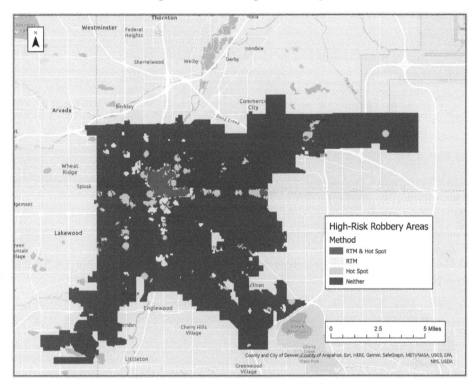

> **Note:** You can change how layers appear in the legend several ways. Changing the name of a layer in the Contents pane automatically changes the name in the legend. The layer properties visible in the legend (for example, layer name, heading, and label) can be adjusted in the Legend Items section of the Legend pane.

Insert a title

1 On the Insert tab, in the Graphics and Text group, click the Rectangle Text button.

2 Drag your pointer to draw a custom-sized title at the top of the map.

3 In the Format Text pane, click the Options button. For Text, type **Denver, CO: High-Risk Robbery Designation.**

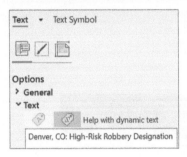

4 Click the Text Symbol tab.

5 Under Appearance, for Size, select 28 pt.

6 Under Position, next to Horizontal Alignment, click the justify text button (second icon from left).

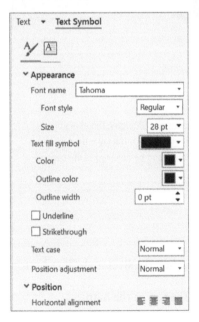

7 At the top of the Format Text pane, click the Text down arrow, and select Background.

8 For Color, set the Symbol color to White.

9 At the top of the Format Text pane, click the Text down arrow, and select Border.

10 For Color, set the Symbol color to Black, and click Apply.

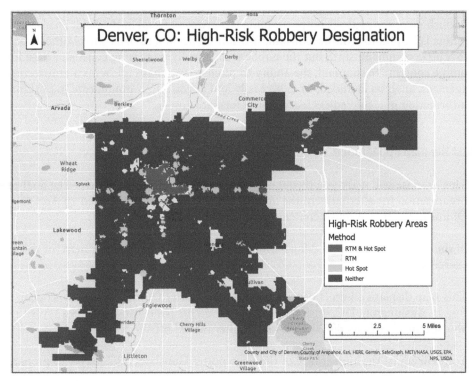

The map is ready to be exported.

Export the layout view to a stand-alone file

1 On the Share tab, in the Export group, click Export.

2 In the Layout pane, for File Type, select PDF.

3 For Name, click the browse button, and browse to C:\ModernPolicing\Chapter9\.

4 Save the file as **Denver_Robbery**, and click Export.

The PDF file of the map appears as a stand-alone file.

5 Browse to your file and open it.

9c

Exercise 9d: Export a web map

Publishing crime analysis outputs online is a growing trend in the field. Sharing maps online is helpful for routine consumers inside an agency, such as patrol officers or executive command staff, but the public is an equally increasing consumer base for online content. To that end, the ability to publish web maps allows law enforcement agencies, analysts included, to more easily and effectively communicate important information in a convenient, easy-to-use format.

Adjust the map appearance

This exercise will walk you through the process to publish a web map. The analyst's organization (or alternatively, the analyst himself or herself) will need an ArcGIS Online account to publish content online, including web maps.

1 In the Catalog pane, in the Maps folder, click Web Map to open the map for this exercise.

This map displays a kernel density map of Shotspotter Gunshot detections occurring in Washington, DC, during 2018. The Contents pane also contains a layer titled Shotspotter Gunshots 80/20 with its visibility turned off.

You will adjust the appearance of the Shotspotter Gunshots 80/20 layer to prepare the map for sharing.

2 In the Contents pane, turn on the visibility for Shotspotter Gunshots 80/20.

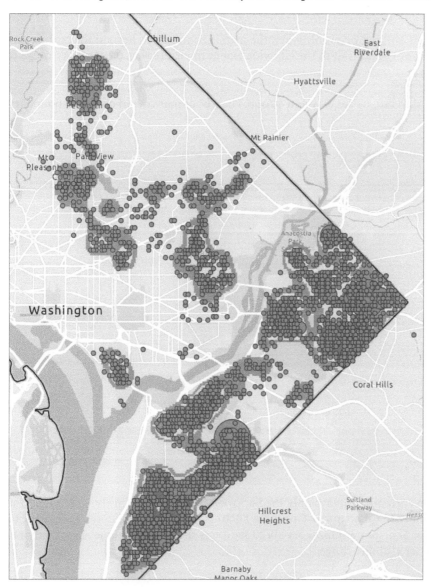

This layer is the output of an 80–20 analysis conducted on the 2018 Shotspotter Gunshot incidents. Displaying these points completely obstructs the visibility of the kernel density layer.

To optimize map visibility at various scales, you will edit the Shotspotter Gunshots 80/20 layer so that it is not visible at the citywide extent. Given the magnitude of points, this layer will be visible only when zoomed closely at a hot spot level.

3 On the Map tab, in the Navigate group, click Bookmarks, and click Eastern Hot Spots.

4 On the Appearance tab, in the Visibility Range group, for Out Beyond, select <Current>.

The points will not be visible when the map is zoomed out beyond the current scale. You will zoom back out to the citywide extent to ensure that the visibility range is set correctly.

5 On the Map tab, in the Navigate group, click Bookmarks, and click DC Boundary.

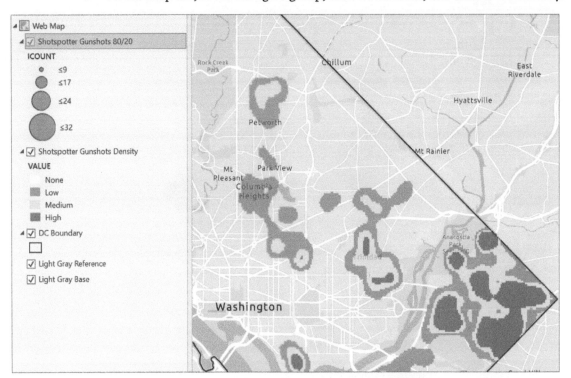

The map currently displays only the Shotspotter Gunshots Density layer even though the visibility for Shotspotter Gunshots 80/20 is turned on. The gray check mark to the left of Shotspotter Gunshots 80/20 indicates that the map scale is currently outside the set visibility range.

Export the web map

1 On the Share tab, in the Share As group, click Web Map.

2 In the Share As Web Map pane, for Name, type **Shotspotter Density**.

3 For Summary, type **This map displays 2018 Shotspotter Gunshot hot spots and 80–20 incident points in Washington, DC.**

4   For Tags, type the following terms, using commas to separate them:
Shotspotter, gunshots, kernel density

5   For Select A Configuration, select Copy All Data: Exploratory.

> **Note:** This is the default setting for sharing web maps. Other options enable you to allow or restrict some settings within your published map, such as data editing by viewers.

6   Under Share With, check Everyone to make the map accessible to users in all groups to which you belong.

7   Click Analyze to determine whether there are any data problems in your map.

8   Click Share after your web map is free of errors.

> **Note:** The map projection must be set to WGS (sphere) for your data to be shared as a web map. Sharing data as a web map may take only a handful of seconds to more than a minute depending on the amount of data in your web map and your network connection speed and settings.

9d

Your new web map can be viewed by clicking Manage The Web Map at the bottom of the pane.

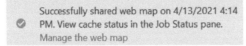

Successfully shared web map on 4/13/2021 4:14 PM. View cache status in the Job Status pane.
Manage the web map

Clicking this link takes you to the item page for your web map in your ArcGIS Online organizational account. You can click Open In Map Viewer to view your map.

Selecting Open In Map Viewer takes you to the finished product your viewers will see. Take a moment to explore the map, observing that it acts as a "mini-GIS" in that you can pan or zoom the map, turn data layers on or off, and select data points to view attribute data. Note that the gunshot points do not appear until the visibility threshold is crossed, similar to the ArcGIS Pro project.

9 In the table of contents, click Share Map.

The Share window appears, allowing you to control the sharing settings for the web map. Here, you can share the map with any ArcGIS groups to which you belong. You can create a custom hyperlink to send to viewers and post the map to social media.

It is important to remember that publishing a web map may be only the first step when sharing content online. ArcGIS StoryMaps℠ stories and web apps are additional ways to bring data and maps to life in a more interactive online environment. As the name implies, StoryMaps stories can best be described as interactive online stories in which viewers can explore GIS data, view related photographs or other visuals, and read text-based narratives embedded in the stories. Web apps, on the other hand, tend to be more data-driven. They allow for more customized viewing and querying of data and GIS maps with tools made available to viewers. Web apps often include useful charts or other metrics-based visuals to further increase understanding of maps and associated data. Creating a web app for community groups or a city council is a suitable use case for such a tool. As powerful as stories and web apps are, however, they rest outside the immediate scope of this book. Entire books and courses can be written about stories and web apps on their own, owing to their immense power and capacity to share knowledge with viewers.

> **Note:** Readers interested in learning more about ArcGIS StoryMaps are encouraged to visit storymaps.arcgis.com.

9d

Summary

The final chapter of this book mirrors what should be the final step of any analyst's efforts: sharing the work that has been completed. The exercises in this chapter taught you how to apply the necessary finishing touches to your work for maximum readability and impact. ArcGIS Pro goes far beyond the standard exporting of static maps to formats such as Adobe Acrobat (PDF) or images (such as PNG or JPG). The robust, interactive environment of ArcGIS Online brings an analyst's data to life. You are encouraged to explore the world of ArcGIS Online, including how to create web apps (such as ArcGIS StoryMaps stories) or use other ArcGIS tools such as ArcGIS Dashboards to combine GIS and tabular data in an easy-to-use format for viewers.

To learn more about free, supported, ready-to-use web app configurations for law enforcement, visit https://doc.arcgis.com/en/arcgis-solutions/industries/law-enforcement.htm.

Customer examples include these dashboards:
- Seattle Police Department
  - https://seattlecitygis.maps.arcgis.com/apps/MapSeries/index.html?appid=94c31b66facc438b95d95a6cb6a0ff2e
- Rochester Police Department
  - www.arcgis.com/apps/opsdashboard/index.html#/415ef13fe15e48ed848b66c52b915efb
- St. Charles Police Department
  - https://policedepartment-stcharles.opendata.arcgis.com/pages/maps
- Long Beach Police Department
  - https://longbeachca.maps.arcgis.com/apps/webappviewer/index.html?id=da05d12e09104e598b6ebcace59f2bba

9d

References

Agnew, Robert. 2011. "Crime and Time: The Temporal Patterning of Causal Variables." *Theoretical Criminology* 15 (2): 115–140.

Andresen, Martin A., and Olivia K. Ha. 2020. "Spatially Varying Relationships between Immigration Measures and Property Crime Types in Vancouver Census Tracts, 2016." *British Journal of Criminology* 60 (5): 1342–1367.

Andresen, Martin, and Nicolas Malleson. 2011. "Testing the Stability of Crime Patterns: Implications for Theory and Policy." *Journal of Research in Crime and Delinquency* 48 (1): 58–82.

Andresen, Martin, Kathryn Wuschke, J. Bryan Kinney, Patricia Brantingham, and Paul J. Brantingham. 2009. "Cartograms, Crime and Location Quotients." *Crime Patterns and Analysis* 2 (1): 31–46.

Anselin, Luc. 1988. *Spatial Econometrics: Methods and Models*. Dordrecht, Netherlands: Kluwer Academic.

Becker, Jacob H. 2019. "Within-Neighborhood Dynamics: Disadvantage, Collective Efficacy, and Homicide Rates in Chicago." *Social Problems* 66 (3): 428–447.

Braga, Anthony, Brandon Turchan, Andrew Papachristos, and David Hureau. 2019. "Hot Spots Policing and Crime Reduction: An Update of an Ongoing Systematic Review and Meta-analysis." *Journal of Experimental Criminology* 15 (3): 289–311.

Braga, Anthony, David Hureau, and Andrew Papachristos. 2011. "The Relevance of Micro Places to Citywide Robbery Trends: A Longitudinal Analysis of Robbery Incidents at Street Corners and Block Faces in Boston." *Journal of Research in Crime and Delinquency* 48 (1): 7–32.

Brantingham, Patricia, and Paul Brantingham. 1998. "Mapping Crime for Analytic Purposes: Location Quotients, Counts and Rates." In *Crime Mapping and Crime Prevention—Crime Prevention Studies, Volume 8*, edited by D. Weisburd and T. McEwen, 263–288.

Brooks, Leah. 2008. "Volunteering to Be Taxed: Business Improvement Districts and the Extra-governmental Provision of Public Safety." *Journal of Public Economics* 92: 388–406.

Buerger, Michael E., and Lorraine Green Mazerolle. 1998. "Third-Party Policing: A Theoretical Analysis of an Emerging Trend." *Justice Quarterly* 15 (2): 301–328.

Cameron, Michael P., William Cochrane, Craig Gordon, and Michael Livingston. 2016. "Alcohol Outlet Density and Violence: A Geographically Weighted Regression Approach." *Drug and Alcohol Review* 35 (3): 280–288.

Caplan, Joel, Leslie Kennedy, and Joel Miller. 2011. "Risk Terrain Modeling: Brokering Criminological Theory and GIS Methods for Crime Forecasting." *Justice Quarterly* 28 (2): 360–381.

Chainey, Spencer, and Jerry Ratcliffe. 2005. *GIS and Crime Mapping*. London: Wiley.

Clarke, Ronald, and John Eck. 2005. *Crime Analysis for Problem Solvers in 60 Small Steps*. Washington, DC: US Department of Justice, Office of Community Oriented Policing Services.

Clarke, Ronald, and Patricia Mayhew, eds. 1980. *Designing Out Crime*. London: Home Office Research Unit, Her Majesty's Stationery Office.

Connealy, Nathan, and Eric Piza. 2019. "Risk Factor and High-Risk Place Variations across Different Robbery Targets in Denver, Colorado." *Journal of Criminal Justice* 60: 47–56.

Cook, Philip, and John MacDonald. 2011. "Public Safety through Private Action: An Economic Assessment of BIDS." *The Economic Journal* 121: 445–462.

Corsaro, Nicholas, Eleazer D. Hunt, Natalie Kroovand Hipple, and Edmund F. McGarrell. 2012. "The Impact of Drug Market Pulling Levers Policing on Neighborhood Violence: An Evaluation of the High Point Drug Market Intervention." *Criminology and Public Policy* 11 (2): 167–199.

Drawve, Grant, Shaun A. Thomas, and Jeffery T. Walker. 2016. "Bringing the Physical Environment Back into Neighborhood Research: The Utility of RTM for Developing an Aggregate Neighborhood Risk of Crime Measure." *Journal of Criminal Justice* 44: 21–29.

Eck, John E. 2017. "Some Solutions to the Evidence-Based Crime Prevention Problem." In *Advances in Evidence-Based Policing*, edited by J. Knutsson and L. Tompson, 45–64. London: Routledge Press.

Goldstein, Herman. 1979. "Improving Policing: A Problem-Oriented Approach." *Crime & Delinquency* 25 (2): 236–258.

Graif, Corina, and Robert J. Sampson. 2009. "Spatial Heterogeneity in the Effects of Immigration and Diversity on Neighborhood Homicide Rates." *Homicide Studies* 13 (3): 242–260.

Haberman, Cory, and Wendy Stiver. 2019. "The Dayton Foot Patrol Program: An Evaluation of Hot Spots Foot Patrols in a Central Business District." *Police Quarterly* 22 (3): 247–277.

Haberman, Cory. 2017. "Overlapping Hot Spots? Examination of Spatial Heterogeneity of Hot Spots of Different Crime Types." *Criminology & Public Policy* 16 (2): 633–660.

Harries, Keith D. 1999. *Mapping Crime: Principles and Practice.* Washington, DC: US Department of Justice.

Hoppe, Laura, and Manne Gerell. 2019. "Near-Repeat Burglary Patterns in Malmö: Stability and Change over Time." *European Journal of Criminology* 16 (1): 3–17.

Hoyt, Lorlene. 2004. "Collecting Private Funds for Safer Public Places: An Empirical Examination of the Business Improvement District Concept." *Environment and Planning B: Planning and Design* 31: 367–380.

Johnson, Shane D., Kate J. Bowers, Dan J. Birks, and Ken Pease. 2009. "Predictive Mapping of Crime by ProMap: Accuracy, Units of Analysis, and the Environmental Backcloth." In *Putting Crime in Its Place: Units of Analysis in Geographic Criminology*, edited by D. Weisburd, W. Bernasco, and G. Bruinsma. New York: Springer.

Johnson, Shane D., Nick Tilley, and Kate J. Bowers. 2015. "Introducing EMMIE: An Evidence Rating Scale to Encourage Mixed-Method Crime Prevention Synthesis Reviews." *Journal of Experimental Criminology* 11 (3): 459–473.

Kennedy, Leslie, Joel Caplan, and Eric Piza. 2018. *Risk-Based Policing: Evidence-Based Crime Prevention with Big Data and Spatial Analytics.* Oakland, CA: University of California Press.

Kennedy, David M. 1997. "Pulling Levers: Chronic Offenders, High-Crime Settings, and a Theory of Prevention." *Valparaiso University Law Review* 31: 449–484.

Lim, Hyungjin, and Pamela Wilcox. 2017. "Crime Reduction Effects of Open-Street CCTV: Conditionality Considerations." *Justice Quarterly* 34 (4): 597–626.

Lum, Cynthia, and Christopher S. Koper. 2017. *Evidence-Based Policing: Translating Research into Practice.* Oxford, UK: Oxford University Press.

Lum, Cynthia, Christopher S. Koper, and Cody W. Telep. 2011. "The Evidence-Based Policing Matrix." *Journal of Experimental Criminology* 7 (1): 3–26.

MacDonald, John, Ricky N. Bluthenthal, Daniela Golinelli, Aaron Kofner, Robert J. Stokes, Amber Sehgal, Terry Fain, and Leo Beletsky. 2009. *Neighborhood Effects on Crime and Youth Violence. The Role of Business Improvement Districts in Los Angeles.* Santa Monica, CA: RAND Corp.

Mastrofki, Stephen D., and James J. Willis. 2011. "Police Organization." In *The Oxford Handbook of Crime and Criminal Justice*, edited by M. Tonry, 479–508. Oxford, UK: Oxford University Press.

McLean, Sarah J., Robert E. Worden, and Moonsun Kim. 2013. "Here's Looking at You: An Evaluation of Public CCTV Cameras and Their Effects on Crime and Disorder." *Criminal Justice Review* 38 (3): 303–334.

Miller, Wilbur R. 1977. *Cops and Bobbies: Police Authority in New York and London, 1830–1870*. Chicago: University of Chicago Press.

Piza, Eric. 2019. *Police Technologies for Place-Based Crime Prevention. Integrating Risk Terrain Modeling for Actionable Intel*. Issues in Spatial Analysis Series, vol. 1. Newark, NJ: Rutgers Center on Public Security.

Piza, Eric. 2018. "The Crime Prevention Effect of CCTV in Public Places: A Propensity Score Analysis." *Journal of Crime and Justice* 41 (1): 14–30.

Piza, Eric, Andrew Wheeler, Nathan Connealy, and Shun Feng. 2020. "The Crime Control Effects of a Police Sub-Station within a Business Improvement District: A Quasi-experimental Synthetic Control Evaluation." *Criminology & Public Policy* 19 (2): 653–684.

Piza, Eric, Brandon Welsh, David Farrington, and Amanda Thomas. 2019. "CCTV Surveillance for Crime Prevention: A 40-Year Systematic Review with Meta-analysis." *Criminology & Public Policy* 18 (1): 135–159.

Piza, Eric, and Andrew Gilchrist. 2018. "Measuring the Effect Heterogeneity of Police Enforcement Actions across Spatial Contexts." *Journal of Criminal Justice* 54: 76–87.

Piza, Eric, Joel Caplan, and Leslie Kennedy. 2014a. "Analyzing the Influence of Micro-level Factors on CCTV Camera Effect." *Journal of Quantitative Criminology* 30 (2): 237–264.

Piza, Eric, Joel Caplan, and Leslie Kennedy. 2014b. "Is the Punishment More Certain? An Analysis of CCTV Detections and Enforcement." *Justice Quarterly* 31 (6): 1015–1043.

Piza, Eric, Joel Caplan, Leslie Kennedy, and Andrew Gilchrist. 2015. "The Effects of Merging Proactive CCTV Monitoring with Directed Police Patrol: A Randomized Controlled Trial." *Journal of Experimental Criminology* 11 (1): 43–69.

Piza, Eric, and Shun Feng. 2017. "The Current and Potential Role of Crime Analysts in Evaluations of Police Interventions: Results from a Survey of the International Association of Crime Analysts." *Police Quarterly* 20 (4): 339–366.

Piza, Eric, and Brian O'Hara. 2014. "Saturation Foot Patrol in a High-Violence Area: A Quasi-experimental Evaluation." *Justice Quarterly* 31 (4): 693–718.

Ratcliffe, Jerry, Travis Taniguchi, Elizabeth Groff, and Jennifer Wood. 2011. "The Philadelphia Foot Patrol Experiment: A Randomized Controlled Trial of Police Patrol Effectiveness in Violent Crime Hot Spots." *Criminology* 49 (3): 795–831.

Ratcliffe, Jerry. 2010. "Crime Mapping: Spatial and Temporal Challenges." In *Handbook of Quantitative Criminology*, edited by A. Piquero and D. Weisburd. New York: Springer.

Ratcliffe, Jerry, and Elizabeth Groff. 2018. "A Longitudinal Quasi-experimental Study of Violence and Disorder Impacts of Urban CCTV Camera Clusters." *Criminal Justice Review*. https://doi.org/10.1177/0734016818811917.

Ratcliffe, Jerry, Matthew Lattanzio, George Kikuchi, and Kevin Thomas. 2019. "A Partially Randomized Field Experiment on the Effect of an Acoustic Gunshot Detection System on Police Incident Reports." *Journal of Experimental Criminology* 15 (1): 67–76.

Ratcliffe, Jerry, Travis Taniguchi, and Ralph Taylor. 2009. "The Crime Reduction Effects of Public CCTV Cameras: A Multi-method Spatial Approach." *Justice Quarterly* 26 (4): 746–770.

Robbins, Michael W., Jessica Saunders, and Beau Kilmer. 2017. "A Framework for Synthetic Control Methods with High-Dimensional, Micro-level Data: Evaluating a Neighborhood-Specific Crime Intervention." *Journal of the American Statistical Association* 112 (517): 109–126.

Sampson, Robert J., Christopher Winship, and Carly Knight. 2013. "Translating Causal Claims: Principles and Strategies for Policy-Relevant Criminology." *Criminology & Public Policy* 12 (4): 587–616.

Santos, Rachel. 2014. "The Effectiveness of Crime Analysis for Crime Reduction: Cure or Diagnosis?" *Journal of Contemporary Criminal Justice* 30 (2): 147–168.

Saunders, Jessica, Russell Lundberg, Anthony Braga, Greg Ridgeway, and Jeremy Miles. 2015. "A Synthetic Control Approach to Evaluating Place-Based Crime Interventions." *Journal of Quantitative Criminology* 31 (3): 413–434.

Schnell, Cory, Anthony Braga, and Eric Piza. 2017. "The Influence of Community Areas, Neighborhood Clusters, and Street Segments on the Spatial Variability of Violent Crime in Chicago." *Journal of Quantitative Criminology* 33 (3): 469–496.

Sherman, Lawrence. 2011. "Police and Crime Control." In *The Oxford Handbook of Crime and Criminal Justice*, edited by M. Tonry, 509–537. Oxford, UK: Oxford University Press.

Sherman, Lawrence. 1998. *Ideas in American Evidence-Based Policing*. Washington, DC: Police Foundation.

Sherman, Lawrence, Patrick Gartin, and Michael Buerger. 1989. "Hot Spots of Predatory Crime: Routine Activities and the Criminology of Place." *Criminology* 27 (1): 27–55.

Skogan, Wesley. 2019. "Community Policing." In *Police Innovation: Contrasting Perspectives,* second edition, by D. L. Weisburd and A. A. Braga, 27–44. New York, NY: Cambridge University Press.

Skogan, Wesley. 2018. "The Commission and the Police." *Criminology and Public Policy* 17 (2): 379–396.

Skogan, Wesley, and Kathleen Frydl. 2004. "Fairness and Effectiveness in Policing: The Evidence." *Committee to Review Research on Police Policy and Practices— Committee on Law and Justice, Divison of Behavioral and Social Sciences and Education.* Washington, DC: The National Academies Press.

Sparrow, Malcolm. 2011. *Governing Science. New Perspectives in Policing Series.* Cambridge, MA: Harvard Kennedy School of Government. Washington, DC: National Institute of Justice.

Sparrow, Malcolm. 2016. *Handcuffed: What Holds Policing Back, and the Keys to Reform.* Washington, DC: Brookings Institution Press.

Steenbeek Wouter, and David Weisburd. 2016 "Where the Action Is in Crime: An Examination of Variability of Crime across Spatial Units in The Hague, 2001–2009." *Journal of Quantitative Criminology* 32 (3): 449–469.

Telep, Cody, Renee Mitchell, and David Weisburd. 2014. "How Much Time Should the Police Spend at Crime Hot Spots? Answers from a Police Agency Directed Randomized Field Trial in Sacramento, California." *Justice Quarterly* 31 (5): 37–41.

Weisburd, David, and Anthony Braga. 2019. "Introduction: Understanding Police Innovation." In *Police Innovation: Contrasting Perspectives*, edited by D. Weisburd and A. Braga, 1–26. Cambridge, UK: Cambridge University Press.

Weisburd, David, and John Eck. 2004. "What Can Police Do to Reduce Crime, Disorder, and Fear?" *The Annals of the American Academy of Political and Social Science* 593 (1): 42–65.

Weisburd, David. 2015. "The Law of Crime Concentration and the Criminology of Place." *Criminology* 53 (2): 133–157.

Weisburd, David. 2011. "Shifting Crime and Justice Resources from Prisons to Police, Shifting Police from People to Places." *Criminology & Public Policy* 10 (1): 153–164.

Weisburd, David. 2008. *Place-Based Policing: Ideas in Policing Series.* Washington, DC: Police Foundation.

Weisburd, David, Wim Bernasco, and Gerben Bruinsma, eds. 2009. *Putting Crime in Its Place: Units of Analysis in Geographic Criminology.* New York: Springer.

Wheeler, Andrew P., and Wouter Steenbeek. 2020. "Mapping the Risk Terrain for Crime Using Machine Learning." *Journal of Quantitative Criminology* https://doi.org/10.1007/s10940-020-09457-7.

White, Matthew B. 2008. *Enhancing the Problem-Solving Capacity of Crime Analysis Units: Problem-Oriented Guides for Police, Problem-Solving Tools Series, No. 9.* Washington, DC: US Department of Justice, Office of Community Oriented Policing Services.

Wilson, Orlando W. 1963. *Police Administration.* New York: McGraw-Hill.

Wilson, James, and George Kelling. 1982. "Broken Windows: The Police and Neighborhood Safety." *The Atlantic Monthly*, March 1982: 29-38.